595.79   Frisch, Karl von

Bees: their vision,
chemical senses,
and language

22067

BEES

Their Vision, Chemical Senses, and Language

BEES

Their Vision, Chemical Senses, and Language

REVISED EDITION

Karl von Frisch

Professor Emeritus of Zoology
in the University of Munich

Cornell University Press

ITHACA AND LONDON

Revised edition first published 1971 by Cornell University Press. Published in the United Kingdom by Cornell University Press Ltd., 2–4 Brook Street, London W1Y 1AA.

Second printing 1972

International Standard Book Number 0-8014-0628-5
Library of Congress Catalog Card Number 71-148718

PRINTED IN THE UNITED STATES OF AMERICA
BY VAIL-BALLOU PRESS, INC.

Contents

Foreword

Karl von Frisch is known throughout the world to both biologists and beekeepers for his discoveries of remarkable sensory capacities and behavior patterns in bees and other lower animals. In this book he reviews these scientific achievements in a straightforward account that requires for its enjoyment neither technical background nor undue effort on the part of the reader. Anyone who has kept a hive of bees has been perplexed at times by the fickle behavior of his charges. In these pages many of the puzzling habits of honeybees are lucidly explained and set into proper perspective as keystones of an elaborate social structure whose smooth functioning has long been the envy of philosophers. So fundamental are certain of the author's discoveries that their impact will surely be felt not only in apiculture and zoology but wherever animal behavior and the mechanism of sense organs are under serious consideration.

In the course of a distinguished scientific career von Frisch has carried out thorough and rigorous investigations of a wide variety of biological problems, including the nature of the pigments in the skins of fish, the color changes of animals, the hearing of fish, the chemical senses of both fish and insects, the vision of insects, and the means of communication employed by bees. His book *Du und das Leben* presents elementary biology so agreeably that it has been widely read in Austria and Germany; it is also well known in English translation as *Man and the Living World*. Born and educated in Vienna, he has been a professor at the universities of Rostock, Breslau, Munich, and Graz. For many years he directed the Zoological Institute at Munich. His work has been widely known and highly respected by professional scientists. Fortunately several of his books, including his *Dance Language and Orientation of Bees*, are now available in English.

The present book begins with a discussion of the cooperative relationship between bees and flowers whereby insects obtain food and the plants achieve the transport of their pollen. We must face the fact that so much has been written about this subject, and with such a surfeit of melodrama, that most of us tend to link it with romantic poetry or nursery tales. One must make a certain effort to clear one's eyes and give fresh, unbiased attention to this matter of the bees and flowers. But here the effort will be well rewarded, for the reader will find that the scientific facts are more intriguing than any artificial embroidery upon them. The third chapter, in particular, leads us to the surprising conclusion that bees manage to communicate to each other precise information about the location of food. The "language" of bees does not employ words, but serves nonetheless to convey complex information and even seems to involve something analogous to map reading.

In recent decades biologists have grown reluctant to credit any claim that the reactions of lower animals attain a high degree of complexity, or what one might be tempted to call intelligence. They recall cases where what was sincerely believed to be intelligent behavior in animals turned out on thorough study to have a much simpler explanation, which did not require us to ascribe higher mental faculties to creatures with much simpler nervous systems than our own. Hence many may feel skeptical at first concerning the conclusions reached in Chapter 3, since their wholehearted acceptance involves a considerable revision of current scientific attitudes. As von Frisch phrased the matter himself on one occasion, "No competent scientist *ought* to believe these things on first hearing."

Von Frisch does not present us with vague or mystical speculations but rather with phenomena which, however astonishing they may be, are nonetheless concrete and readily observed. Now that he has told us what to look for in the seething turmoil of bees creeping over the honeycomb, now that his insight has made order where there seemed to be utter chaos, anyone with a little patience and a hive of bees can test the principal conclusions for himself.

Indeed, independent confirmation was soon accomplished in the United States, in England, and on the continent of Europe. For example, Dr. W. H. Thorpe of Cambridge University wrote in *Nature* (164 [1949]: 11–14):

Through the kindness . . . of Professor von Frisch I was able . . . to perform with him and for myself . . . "repeats" of certain of the most crucial experiments. . . . My observations covered all the main phases of the work, and I was able to make my own estimates of the efficiency of the indication of distance and direction of food sources. . . . This memorable experience . . . enabled me to resolve to my own satisfaction some of those doubts and difficulties that come to mind on first reading the work, and convinced me of the soundness of the conclusions as a whole.

I confess without embarrassment that until I performed these simple experiments myself, I too retained a residue of skepticism. But a few weeks' work with an observation beehive and a colony of bees (loaned for the purpose by the Department of Entomology of the New York State College of Agriculture) led me to the same degree of conviction as that which Thorpe reports. While certain details remain unclear, and while much additional work must be done before the dances and the "language" of the bees are fully understood, the important basic facts described in Chapter 3 appear to be established beyond serious question. Recently doubt has been expressed by Wenner and Johnson that information is actually conveyed from one bee to another by the dances discovered by von Frisch (*Science*, 155 [1967]: 844–849, and 158 [1967]: 1076–1077). But these objections have been disposed of quite adequately by von Frisch (*Science*, 158 [1967]: 1072–1076), Esch and Bastian (*Zeitschrift für vergleichende Physiologie*, 68 [1970]: 175–181), Gould, Henerey, and MacLeod (*Science*, 169 [1970]: 544–554), and Lindauer (*Amer. Naturalist*, 105 [1971]: 89–96).

While the conclusions of this book are of basic importance to biological science and are truly revolutionary in the special field of animal behavior, the phenomena themselves display an intrinsic simplicity that is characteristic of many classic experiments. Hence they provide for the general reader a rare opportunity to appreciate the mode of thinking and the point of view of the critical investigator. The account is condensed, to be sure, with many laborious intervals and false starts left out; but nonetheless as one reads these pages he can feel a real sense of participation in the search for understanding.

Appreciation of a scientist's mode of thinking requires more than a bare scrutiny of phenomena, hypotheses, experiments, and conclusions. The thought and the word are closely linked together; and for this reason an effort has been made to preserve in the printed page something of the pleasing directness and simplicity so apparent in the original lectures. Those who heard the lectures will recognize this flavor; and I believe that many others, a trifle jaded, perhaps, by the conventional jargon of scientific writing, will be refreshed by the straightforward clarity of this account.

In the approximately twenty years since this book was first published, new findings and new ramifications have reinforced all the opinions expressed in my original foreword. For example the fine work of Martin Lindauer, begun in close collaboration with von Frisch and summarized in Lindauer's book *Communication among Social Bees*, has added an important new dimension by showing that bees communicate to other bees not only the location of food sources but also of possible new homes for a swarm. This disposes of simplistic interpretations that link the dancing behavior rigidly to hunger or energy metabolism.

As anticipated by August Krogh (*Scientific Amer-*

ican, August, 1948), the implications of von Frisch's discoveries have been actively discussed both by biologists and philosophers. One of the clearest examples of the interest, and essential skepticism, of the latter is Jonathan Bennett's book *Rationality: An Essay towards an Analysis* (New York: Humanities Press, 1964). Bennett, like Mortimer J. Adler (*The Difference of Man and the Difference It Makes* [New York: World, 1967]), refuses to accept the bee dances as a true language, primarily because there is no evidence of conscious intent on the part of the bees. Since we lack any sources of reliable evidence about conscious intent in other animals, whether they be chimpanzees or honeybees, this criterion seems to me at best inconclusive and at worst misleading. A reluctance to become embroiled in metaphysics should not anesthetize our perceptions. Heretical as it may seem to many behavioral scientists, I am willing to entertain the thought that perhaps the bees know what they are doing.

DONALD R. GRIFFIN

The Rockefeller University
and New York Zoological Society

Preface to the Revised Edition

Since the first edition of this book was prepared, twenty years have passed, years of intensive study of the problems under discussion. Consequently I am happy to accept the publisher's suggestion that the text be re-examined and brought up to date. An appendix concerning the perception of polarized light had already been added in the second printing. But now it seems more appropriate to give up a literal reproduction of the original lectures and to incorporate all new information in the text. Inasmuch as the new knowledge is based on what was known previously, considerable reduction has been possible. Nine figures have been replaced by better ones, twenty new ones have been added, and two have been eliminated altogether. I wish to thank Dr. Leigh E. Chadwick most sincerely for his excellent translation of the changes and additions to the text.

It would have been easy to expand this little book

into a thick tome. But that was not the object. The volume is intended to remain what it has been hitherto: an easily read introduction to one of the most fascinating areas of biology. It is designed to show the layman what sorts of problems are at issue here, and how they may be solved. I have tried to present intelligibly and to adapt harmoniously to the old text even the more difficult of the new discoveries. Anyone who wishes to ascertain what new knowledge has been accumulated during the last twenty years can judge this from the dates of the references cited.

But knowing the date of individual discoveries does not matter a great deal. It is more important to consider in their entirety the accomplishments of these small insects, the bees. The more deeply one probes here the greater his sense of wonder, and this may perhaps restore to some that reverence for the creative forces of Nature which has unfortunately been lost by so many of today's people.

KARL VON FRISCH

Munich
March 17, 1971

Preface to the First Edition

During the spring of 1949 I made a three-month lecture tour through the United States, accompanied by my wife. The invitation for the trip came first from Cornell University. It was later extended to include sixteen other universities and scientific institutions, and was supported jointly by them and the Rockefeller Foundation. My only regret was that time did not permit me to accept similar invitations from several other universities in the United States and Canada.

This book offers the text of three lectures given at Cornell University, at the American Museum of Natural History, New York City, and at the University of Minnesota; only the third lecture was presented at the remaining universities. It is a pleasure to be able to present the problems discussed in these lectures to a larger circle. Of course it was impossible in three lectures to describe the detailed experimental basis for the conclusions reached after almost forty years of work, much of which was carried out in collaboration with the coworkers and colleagues mentioned in the following pages. Hence a bibliography is included so that the reader can obtain more information about the findings and about our methods.

I am deeply indebted to Dr. D. R. Griffin of Cornell University for the careful planning and advance preparations for our trip, and for assistance with the English of the manuscript. Furthermore, I wish to express my thanks to all the other people whose cooperation made this lecture tour so interesting and useful to us. For many years, we have missed such contacts and the exchange of ideas with foreign countries. Scientific work must be international and cannot prosper if confined in a cage. I have learned more during these three months, in both personal and scientific matters, than in the three previous years at home.

We shall never forget the great kindness and hospitality of everyone we met in your beautiful country.

KARL VON FRISCH

Munich
May 1, 1950

1. The Color Sense of Bees

The honeybee, living in its beehive, is a social insect. In an ordinary beehive there are about sixty thousand bees, but only one is a fully developed female. This is the queen, the only egg-laying insect in the colony. The males or drones are larger, more plump, and a little stupid and lazy. All of the remaining bees are workers (Figure 1). The workers are not able to pro-

Figure 1. Left to right: The queen. The worker bee. The drone. *H.*, head; *Th.*, thorax; *Ab.*, abdomen; *A.*, antenna; *E.*, eye.

duce eggs, under normal circumstances, for their ovaries are small and undeveloped; but in other respects they are females, with the typical instincts of females; they do all the work in the hive. They feed the larvae, they build the honeycomb, they are the charwomen of the hive; and it is only these worker bees that fly out to gather honey and pollen as food for the colony.

The worker bees find the food by visiting flowers.

Figure 2. Left: Grass blossoms as an example of blossoms pollinated by the wind.

Figure 3. Right: The blossoms pollinated by insects are the larger and more conspicuously colored ones.

Here some gather nectar droplets with a high concentration of sugar. Others collect pollen, since they also need protein for the growing larvae. But in taking their food they do not behave like plunderers. They reciprocate and perform a service for the plants by effecting the pollination—flying from one flower to the next and carrying the pollen adhering to their bodies.

It is well known that there are two main types of "flowers" among the higher plants. Many plants have small green blossoms without any scent, and the transfer of pollen is effected by the air (Figure 2). Such plants produce an abundance of pollen, which is spread by the wind and comes by chance to other blossoms of the same species. Other plants have conspicuous, brightly colored blossoms or a striking scent, or both, and it is these that we ordinarily call flowers (Figure 3). Only such flowers produce nectar and are therefore visited by insects, which effect the pollination by flying from one flower to the next (Figure 4). Biologists have long believed that flowers are colored and scented to make them more striking for their insect visitors. In this way the insects can more easily find the flowers and get their food; and the pollination is also assured.

But this view has not been accepted by all biologists. About 1910 a famous ophthalmologist, Professor C. von Hess, performed many experiments on fishes, insects, and other lower animals. He tested them while they were in a positively phototactic condition—that is, under circumstances where they moved into the brightest available light. He found that in a spectrum the animals always collected in the green and the yellow-green region, which is the brightest part of the spectrum for a color-blind human eye. Therefore, von Hess asserted, fishes and invertebrates, and in particular bees, are totally color-blind. If this were true, the colors of flowers would have no biological significance. But I could not believe it, and my skepticism was the first motive which led me to begin my studies of bees about 1910. I tried to find out whether bees have a color sense.

Figure 4. Nectar is produced in the bottom of a flower, so that as they suck it up the visiting insects come in contact with the pollen.

By the scent of a little honey it is possible to attract bees to an experimental table. Here we can feed them on a piece of blue cardboard, for example. They suck up the food and, after carrying it back to the hive, give it to the other bees. The bees return again and again to the rich source of food which they have discovered. We let them do so for some time, and then we take away the

Figure 5. Bees fed previously on a blue card in the middle of the table (*) alight on the clean blue cardboard without food (*left*). They distinguish it from a red cardboard (*right*).

blue card scented with honey and put out two new, clean pieces of cardboard at the site of the former feeding place—on the left a blue card, and on the right a red one. If the bees remember that they found food on blue, and if they are able to distinguish between red and blue, they should now alight on the blue card. This is exactly what happens (Figure 5).

This is an old experiment. It indicates that bees can

distinguish colors, but it does not prove that they have a color sense, or color perception, for these are not always the same. Thus there are totally color-blind men, although they are very rare. They see objects as we would see them in a black-and-white photograph. Yet they can distinguish between red and blue, for red appears very dark to them, and blue much lighter. Hence we cannot learn from the experiment with bees which I have just described whether the bees have distinguished red from blue by color or by shade, as a color-blind man might do. To a totally color-blind man each color appears as a gray of a certain degree of brightness. We do not know what the brightness of our various pieces of colored cardboard may be for a color-blind insect. Therefore we perform the following experiment.

On our table we place a blue card and around it we arrange gray cards of all shades from white to black. On each card we set a little watch glass, but only the glass dish on the blue card contains food (sugar-water). In this way we train the bees to come to the color blue. Since bees have a very good memory for places we frequently change the relative positions of the cards. But the sugar is always placed on the blue card so that in every case the color indicates where food is to be found. After some hours we perform the decisive experiment. The cards and the glass dishes soiled by the bees are taken away. We place on the table a new series of clean cards of different shades of gray, each with an empty glass dish, and somewhere among them we place a clean, blue card provided, like all the others, with an empty glass dish. The bees remember the blue color and alight only on the blue card, distinguishing it from all shades of gray. This means that they have a true color sense.

This type of experiment has been criticized on the ground that the blue cardboard might have a specific scent by which the bees could recognize it. We cannot perceive any odor, but this does not prove that it is

Figure 6. Bees trained to the color blue alight on a clean blue cardboard without food and under glass—distinguishing it from all degrees of gray.

odorless for bees; we must therefore consider the possibility that the bees found the blue cardboard by smell, and not by color. But this is not the case. For we can repeat the experiment with a glass plate lying over the cards; if there were any scent it could not pass through the glass. But the outcome of the experiment is just the same as before (Figure 6).

Training bees to come to food on orange, yellow, green, violet, or purple cardboard gives the same positive result. However, if we try to train bees to find their food on scarlet red, they alight not only on the red cardboard but also on black and on all the dark-gray cards in our arrangement. Thus red and black are the same to the eye of the bee; in other words, bees are red-blind. From these experiments it is clear that bees have a color sense, but that it is not quite the same as that of a normal human being.

To find out something more about the nature of color perception in bees, we modify our experiment. We train bees to find food on blue, and then we put on the table all the *colored* cards we have on hand, but no gray cards at all. The bees seek for blue, but it is surprising to see that they are unable to find it with certainty. They confuse blue cards with violet and purple ones. Furthermore, bees trained to yellow alight not only on yellow cards but on orange and green ones, too (von Frisch, 1915).

In 1927 Professor A. Kühn repeated my training experiments, but instead of colored cards he used a spectrum. He was able to confirm my results: that bees are red-blind, that they can distinguish other colors from all shades of gray, and that they confuse yellow with orange and green, or blue with violet. But by using spectral colors he discovered two new facts: First, he noted that there is a third quality of color in the narrow blue-green region (480–500 mμ [1]). Bees trained to blue-green distinguish it from blue and from yellow. I had overlooked this point because I had no suitable cardboard of this blue-green color. Second, he discovered a fourth quality in ultraviolet light. If bees on the experimental table are fed for some time in ultraviolet light, they alight on every spot irradiated by ultraviolet, even though this light is invisible to us; and they distinguish the ultraviolet from all shades of white or gray. It is a distinct color for the bees.

If we compare the color sense of bees and men, we find that the visible spectrum is shortened for bees in the red but that it is extended into the ultraviolet. In this way the visible region is merely shifted to shorter wave lengths. But a much more important difference seemed to be that the bee apparently sees only four different qualities of color: yellow, blue-green, blue, and ultraviolet (Figure 7).

[1] See Figure 7, legend.

In order to make a closer comparison between the color vision of bees and man, experiments with color mixtures [2] would have been important. At the time

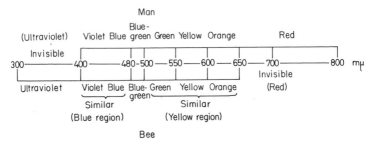

Figure 7. The colors of a spectrum for the human eye (*above*) and for the eye of the bee (*below*). The numbers indicate the wave length of light in millimicrons, mμ (one micron = 0.0001 centimeters or 1/25000 inch). (From von Frisch, *Tanzsprache und Orientierung der Bienen* [Springer-Verlag, 1965], p. 486; *Dance Language and Orientation of Bees* [Harvard University Press, 1967], p. 475.)

we lacked adequate apparatus. Other problems pushed their way into the foreground and so it was thirty years

[2] Here and in what follows color mixing always signifies true physiological mixing (additive mixing), in which, for example, we allow two different spectral colors to work simultaneously on the eye or with the help of a rotating disk present the eye with two colored pigments in such rapid alternation that they are viewed practically at the same time. Subtractive mixing of colors, such as the artist practices, produces different hues. These come into being on the very palette; the initial colors never reach the eye.

before I induced K. Daumer to fill in these gaps and to resume with improved technique the training of bees to colors. Then for the second time we were made aware that improved methods brought better performance in the area of color vision. Daumer (1956) trained the bees individually. Doing so gives more clean-cut results, because the bees disturb one another when they are trained in groups. Customarily they flew through a window into a semidarkened room that held the training cabinet (Figure 8). A lamp inside the cabinet gave out white light, including ultraviolet which

Figure 8. Cabinet for training to color; inside it are the light source and the apparatus for mixing colors. In the top are four star-shaped openings that are covered by glass transparent to ultraviolet and that bear quartz vessels (transparent to ultraviolet) for food and water. After Daumer, 1956.

bees are able to see. The distribution of energy over the several spectral ranges was like that of sunlight. Four star-shaped openings in the lid of the cabinet were covered with glass transparent to ultraviolet. Upon them stood quartz feeding dishes, likewise transparent for ultraviolet (Figure 9). Each star could be illuminated from below with colored light from a narrow region of the spectrum or with a mixture of various spectral colors. After a bee had been trained to a given

Figure 9. Bee at the feeding vessel on a star illuminated from beneath. After Daumer, 1956. (From von Frisch, *Tanzsprache und Orientierung der Bienen* [Springer-Verlag, 1965], p. 484.)

color, or to a mixed illumination, she was tested experimentally for her ability to distinguish this from other colors offered alongside it. In this experimental setup the bees were indeed able to differentiate orange, yellow, and green. Yet these colors are more alike for them than for man. For bees blue-green is a sharply contrasting separate color. Within the blue-violet and ultraviolet ranges bees can recognize several different hues. But they can see more than this. In order to comprehend such matters we require a short lesson in physiological optics.

When white sunlight is beamed through a prism it acquires an orderly arrangement according to the wave length, and the colorful ribbon of the spectrum appears (Figure 10a). If by means of a lens we reunite all the colors, white is found again (Figure 10b). But if we bring together only the ends of the spectrum, namely red, of long wave length, and violet, of short wave length, then there are formed the purple colors, that do not occur in the spectrum (Figure 10c). The same applies to bees. Only here we have to combine those colors that lie at the ends of *their* spectrum, to wit yellow and ultraviolet. Daumer's training experiments showed that the mixture of yellow and ultraviolet likewise looks to the bees like a color, clearly distinct from white but also from all the spectral colors.[3] If with a person we change the proportion in which red and violet are mixed, then we can produce purple in all possible transitions between these two colors and build a

[3] What a bee actually feels we are of course unable to find out. From their behavior we can only judge what things strike them as alike, similar, or different.

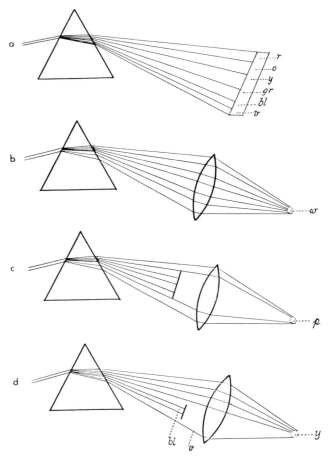

Figure 10. White light sent through a prism is put in order according to wave length and gives a spectrum (*a*). United again by a lens, the light becomes white once more (*b*). The ends of the spectrum give purple when mixed together (*c*). Without the blue rays the light becomes yellow (*d*).

continuous bridge between the two ends of the spectral ribbon, thus forming a complete color circle. The same kind of thing happens when yellow and ultraviolet are mixed to form "bee purple" (Figure 11).

Returning once more to Figure 10, we do another experiment: we screen off none but the blue rays. Then the mixture of all the other rays no longer looks white, but yellow (Figure 10d). By removing a small portion of the spectrum, white has been turned into a color.

Now naturally we are curious to know whether in addition to the colors there is a "white" for bees. The question is more interesting than it seems at first glance. With the human eye the sensation of colorless "white" is produced when all the colors of the spectrum are mingled in a definite proportion—such as occurs in sunlight. When all colors that *bees* can perceive (hence inclusive of ultraviolet) are mixed together, a special kind of light is produced for them too. We give it the name "bee white." For them it is not like any color. Also it does not attract them, as colors do. Only with difficulty can they be trained to it. But if the ultraviolet is taken away, then the light is no longer white for them, but blue-green. All of this can be demonstrated by means of training experiments. Since we do not perceive ultraviolet, removal of the ultraviolet does not change the white for us. Thus, what looks "white" to us may be two different things for bees: with ultraviolet it is "bee white," but without ultraviolet it is a bright blue-green. The white of flowers has these characteristics (Hertz, 1937, 1939; Daumer, 1958).

Via the experiments with color mixtures, some striking areas of agreement between the color vision of man and bees have become evident: the possibility of producing purple colors by combining the ends of the spectrum, or that of making "white" by mixing *all* the spectral colors. Such white can also be produced for the bee, as for man, by combining *pairs* of colors, the "complementary colors" (they are situated opposite one another in the color circle of Figure 11).

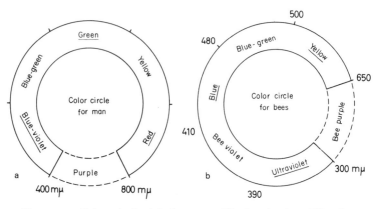

Figure 11. Color circle (*a*) for man; (*b*) for the bee. The three primary colors are underlined; by mixing them the intermediate colors can be produced. Complementary colors are opposite each other in the circles. Somewhat modified, after Daumer. (From von Frisch, *Tanzsprache und Orientierung der Bienen* [Springer-Verlag, 1965], p. 487; *Dance Language and Orientation of Bees* [Harvard University Press, 1967], p. 476.)

By mixing the *three primary colors* (underlined in Figure 11) in the proper proportions, one can produce white, or also any of the colors. As long ago as in the previous century this fact led to the assumption that there are in our retina three different kinds of color-sensitive elements that respond maximally to differing ranges of the spectrum. By means of their collective action we are enabled to see white and all the colors (the theory of trichromatic color vision of T. Young and H. von Helmholtz). That was a hypothesis that just very recently has been confirmed almost simultaneously for both bees and man. It is now possible to insert extremely fine electrodes into individual sensory cells in the bee's eye and to measure the excitation in them. In doing this there have been found three types of sensory cell, with maximal sensitivity in the yellow-green, blue, and ultraviolet, respectively (Autrum and von Zwehl, 1964). A different method succeeded in demonstrating in the human retina three kinds of color-sensitive sensory cells (cones), each adapted to a different spectral range (Brown and Wald, 1964).

From these and additional discoveries it appears that color vision in bees resembles that of man more closely than had been thought, and that its physiological basis is the same. But as a significant difference there remains the fact that the visible spectrum is displaced for bees: cut short at the long wave-length end and extended with the short wave lengths (Figure 7). This fact may leave us well content. For here we see with remarkable clarity that floral colors have arisen as an adaptation to the color sense of those creatures that frequent the flowers. I would like to demonstrate this with a few examples:

Already I have mentioned *one* matter that points in this direction: white flowers absorb ultraviolet light and therefore look like colored (blue-green) flowers to the bees. It is only through their coloration that they become attractive to bees. Additionally, we now understand why scarlet flowers are so rare in Europe. The insects that frequent them are red-blind. In America and Africa there are many scarlet flowers. But long before we knew anything about the color sense of insects people were aware that that kind of red is typical of flowers that are pollinated not by insects but by humming birds and honey birds (see Porsch, 1931). For these birds red is a brilliant color.

Poppies are red flowers that are frequented by bees. This exception has an amazing explanation. Poppies reflect the ultraviolet rays of sunlight. It is possible to show by suitable experiments that bees trained to the blossoms of the poppy are in fact recognizing the reflected ultraviolet light (Daumer, 1958). We cannot perceive this light, and we see only the red. The bees, on the other hand, cannot perceive the red; they see only the ultraviolet. Thus the bees, although they are

Figure 12. Flowers of (*top*) *Erysimum helveticum* (wormseed mustard), (*middle*) *Brassica napus* (rape), (*below*) *Sinapis arvensis* (field mustard), photographed in yellow light (*left*) or in ultraviolet light (*right*). Differences in ultraviolet reflection produce for the eye of the bee clearly distinct colors in the flowers, all of which are of the same uniform yellow to us. After Daumer, 1958. (From von Frisch, *Tanzsprache und Orientierung der Bienen* [Springer-Verlag, 1965], p. 496.)

red-blind, nevertheless see the poppy as a colored flower —for them it has the color ultraviolet.

Other floral colors also are often changed by means of ultraviolet in a way useful to bees. Figure 12 gives an example. Illustrated together are three species of flowers that have the same yellow color and very similar shapes. Since they occur in the same place the bees could easily mix them up. In the left-hand row they have been photographed through a yellow filter, on the right through an ultraviolet filter. One sees that *Erysimum*, the wormseed mustard (*above*), reflects no ultraviolet; *Brassica*, rape (*middle*), reflects some ultraviolet; and *Sinapis*, field mustard (*below*), a great deal. None of the three flowers reflects blue light. Accordingly, only *Erysimum* is yellow for the eye of the bee; the other two species wear "bee purple," and display this in two different hues readily distinguished by bees. Similar situations are known in blue flowers. For bees the presence of ultraviolet augments greatly the multiplicity of floral colors. How important that is becomes evident when we inquire into the biological significance of the colors of flowers.

What is their biological meaning? Biologists have long believed that color plays an important role in making flowers attractive to bees and other insects. This is certainly true for bees flying out in search of new flowers from which to gather food. My collaborator,

Dr. Therese von Oettingen (1949), set up a beehive in a courtyard with a roof of fine-mesh screening, and studied bees that had never before visited flowers and had never had the opportunity to leave the observation room. She displayed colored papers at several points in the room, and at other points she placed scented flowers which could be smelled by the bees but which were covered and could not be seen. A few of the bees were scouts, and these were attracted to both the colors and the scented flowers. From this it is clear that both odors and colors are attractive to bees that are seeking new feeding places.

We were surprised to find that very few bees—only the scouts—paid any attention at all to the flowers or to the colors and scents that were displayed. In forty experiments, lasting forty-five hours altogether, Dr. von Oettingen used hundreds of bees and fifteen different species of flowers; but she found that a group of flowers was visited by only one or two bees per hour (average, 1.27). The maximum was six bees during one hour. The great majority of the worker bees did *not* seek food even though there was very little nourishment available in the hive. As with human beings, pioneers seem to be rare in the beehive. Most individuals prefer to wait for the discoveries of a few scouts in order to find food by following their instructions, as I shall describe in the third chapter.

When bees have found a species of flower with abundant nectar, then they are loyal to this kind of flower for many days. A given individual bee on its foraging trips always visits a particular species of flower. This is of advantage to the bees, which find on all blossoms of the same species the same flower mechanism and save time through being familiar with it. It is also of advantage to the flowers, for their pollination depends on bees coming from other flowers of the same species. Since the bees specialize in certain flowers, they must be able to distinguish between the various species present on their feeding grounds. Hence it is a good thing if different kinds of plants have flower colors as different as possible, that are easily distinguished by the bees.

But colors can be useful to bees in still another way. We often find flowers in which the entrance to the tube containing the nectar is of a different appearance from the remainder of the flower—either a darker or lighter shade or, very often, a different color. It is remarkable that the color difference in nearly every case results in an impressive contrast for the eye of the bee. In German we call such colored spots *Saftmale*—sap spots; in English, "nectar guides" (Figure 13). More than 150 years ago C. K. Sprengel concluded that these colored spots were signposts helping visiting insects to find the nectar. Later biologists were often skeptical about this point. But Sprengel was right. One can test the matter by simple experiment. If we put upon a table a large blue card on which we have placed a small yellow spot, this spot is attractive to bees and they prefer to alight on such a little spot to seek for food. Daumer (1958) found that, on the flowers too, very small nectar guides sufficed to show the bees where to insert the

Figure 13. The blue flowers of the forget-me-not (*Myosotis*) have a yellow ring (sap spot) around the entrance to the nectar.

trunk. When they come upon the nectar guide they often twitch the head downwards and extend the proboscis. This is an innate response. It was also seen in a colony that had been kept in a closed chamber for several years and had never seen flowers before the experiment. It was not perception of the nectar that caused proboscis extension. By doing a little operation on the flowers Daumer displaced the nectar guide to the outer margin. Now the bees, who ran hunting about on the flower, did not give the proboscis response in its center where the nectar was, but at its edge, to which the nectar guide had been shifted.

We too can see nectar guides on many flower species. But such things are twice as frequent for bees as for us. There are nectar guides that are invisible to us but very apparent to them. Here again is demonstrated the adaptation of the floral colors to the color sense of their visitors. An example is shown in Figure 14. The cinque-

Figure 14. Potentilla reptans (cinquefoil), photographed through yellow (*left*), blue (*middle*), and ultraviolet (*right*) filters. Below is the gray scale used for photometric comparison. The leaves, photographed at the same time, give only weak and almost uniform reflections in the bee's three primary color ranges, so that they are almost colorless for bees. Through the ultraviolet filter there is revealed a special nectar guide, invisible to us. After Daumer, 1958. (From von Frisch, *Tanzsprache und Orientierung der Bienen* [Springer-Verlag, 1965], p. 498.)

foil (*Potentilla reptans*) flower, pure yellow for us, has been photographed through yellow, blue, and ultra-violet color filters. The color filters had been prepared so as to correspond with the three primary colors of the bee's eye. One can see that the flower reflects yellow strongly, blue not at all, and ultraviolet only with the outer portions of the petals. Consequently, for bees the flower has a purple color with a yellow nectar guide in the center.

In this same illustration we may notice something else also, and this is extraordinarily interesting. A few of the plant's green leaves have been photographed too. In all three of the bee's primary colors they reflect the light weakly and fairly equally, being a tiny bit stronger in the yellow. The same thing was confirmed with other foliage. This means that the foliage, which is green for us, is for bees an almost colorless gray, with a weak yellowish tinge. The colorful flowers must stand out very effectively against this neutral background.

One can see that the colors of flowers have been developed as an adaptation to the color sense of their visitors. It is evident that they are not designed for the human eye. But this should not prevent us from delighting in their beauty.

In Austria, and in Germany too, beekeepers put their beehives all together, one beneath another, to form a beehouse. In a large house with many hives, it is a little difficult for the homing bees to find their own hive; as a matter of fact, they very often shift from one hive to another. For the most part this does not matter, because bees carrying honey are welcomed everywhere. But sometimes a struggle occurs, and intruding bees may even be stung to death. A queen returning from an orientation flight or from a mating flight is in particular danger; if she alights on the wrong hive she is killed. In our country, therefore, we often see hives painted with different colors to help the bees recognize their own homes. But all beekeepers do not agree that this coloring is useful.

To study this question I placed a swarm of bees in one of a row of empty white hives and covered the front of this hive with a blue sheet of metal. On the right stood a hive that I covered with a yellow sheet, and on the left a white hive. After some days I wished to change the colors. But if I had merely interchanged the colored sheets, and if the bees had then flown to the wrong hive, I would not have known whether they were following the blue color or the scent of bees adhering to the blue sheet. Therefore the back side of the blue sheet was painted yellow, and the back of the yellow one blue. In this way I could change the colors by turning each sheet around, without moving it to a different hive. After thus reversing both metal sheets I saw many bees flying to the empty hive, which was now blue. Other bees hesitated, and after some time flew to the correct hive, despite the yellow sheet (Figure 15).

I decided that perhaps the bees were paying attention not only to their own hive but to the neighboring hives as well. In this case my experiment must have caused confusion. The bees had been accustomed to fly to a blue hive with a yellow one on the right, and a white

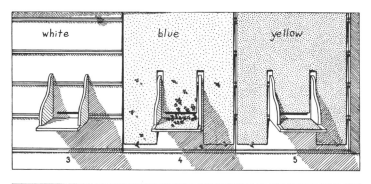

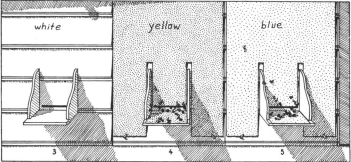

Figure 15. Upper: There is a colony of bees in the blue hive (4) situated between the white and the yellow ones, which are both empty. *Lower:* By reversing the blue sheet, the back of which was painted yellow, and also reversing the yellow sheet, which is blue on the back, the colors of the two hives have been interchanged. Many bees enter the wrong hive (5), following the color to which they are accustomed.

hive on the left. Now they saw a blue hive with a yellow one on the left and a white one on the right. Here was a different arrangement; and perhaps this explained why some of the bees hesitated and finally found the correct hive by the sense of smell. I therefore repeated the experiment in another way. I reversed the blue

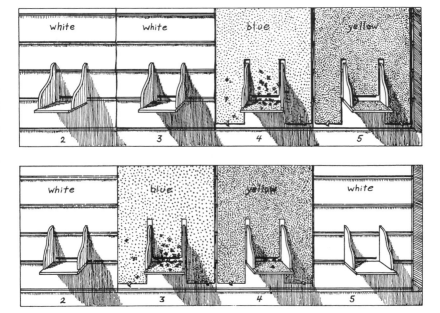

Figure 16. Upper: The initial color pattern of the blue hive (4), which is occupied by the bees, and the hives adjacent to it. *Lower:* The color of hive 4 has been changed to yellow by reversing its metal sheet, and the yellow sheet covering empty hive 5 has been turned and transferred to hive 3, which now appears blue. All of the homing bees which came from hive 4 now enter the empty hive (3) which is blue.

sheet (so that it appeared yellow), but I shifted the re-versed yellow sheet (which then appeared blue) from the right side to the left. The relative positions of the colors were now unchanged; the bees still found a blue hive with a yellow one on the right, and there was still a white hive to the left of the blue one (since all the hives in the row were white except those marked with my metal sheets). Under these circumstances every bee, without a single exception, alighted at the wrong hive and entered it even though this hive was quite empty (Figure 16). We can see from this experiment that a painted beehive is actually a very good signpost to help bees recognize their own hive and distinguish it from others nearby (von Frisch, 1915).

But why do beekeepers disagree with one another about the effectiveness of colored hives? Simply because they have not considered the nature of the color sense of bees, as the flowers seem to have done. We often see hives stacked one beneath the other, one painted red and the next one black. This forms a contrast for the human eye, but the two look just alike to bees. For bees blue, yellow, black, and white are readily distinguishable. Where beekeepers must use the same color repeatedly they should change the color patterns formed by adjacent hives, because bees pay attention to the color of the neighboring hives as well as to their own (Figure 17). Furthermore, when a hive is to be painted white, it is important to choose the proper white. Zinc white seems to be suitable for this purpose because it absorbs ultraviolet and therefore looks blue-green to the bees. Lead white reflects ultraviolet and thus it is really white for bees, and hence less striking than a true color. Instead of black, scarlet can be used. This is the same as black to the bees and makes things more colorful for us.

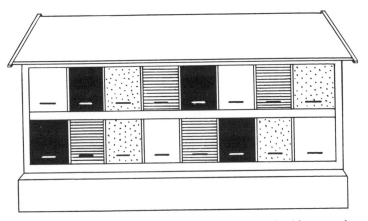

Figure 17. An example showing how to paint beehives so that it becomes as easy as possible for the bees to find their home. Colors used are white, blue, yellow, and black.

Beekeepers in the United States do not use beehouses, but place their beehives on the ground at considerable distances from one another, in the open. I have sometimes seen many beehives placed in a meadow without any striking landmarks around them. It is probable that here too the bees may often confuse their home with other hives. I believe that painting the hives according to the principles explained above would be as useful to American beekeepers as to their European colleagues.

But let us return to the flowers. Their color is a good indicator for foraging bees to find out the desired species, and makes possible their constancy toward a given kind of flower. Is the flower's shape of help in this? In order to discover that, I trained bees to recognize a cer-

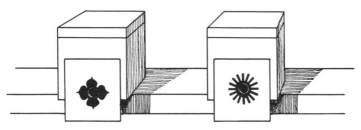

Figure 18. Bees can easily distinguish shapes like these.

tain pattern of colored paper pasted around the entrance of a cardboard box containing sugar-water. To other boxes, without food, I attached papers of the same color but cut in different patterns. Using the patterns shown in Figure 18, I succeeded within a short time in training bees to go to the box marked with one of these figures. For example, in one experiment food was made available for some time in the box marked with the radiate pattern; then I took away this box and placed on the experimental table several new, clean boxes not yet touched by bees and marked with pieces of colored paper cut in the two patterns of Figure 18. Many of the bees entered the boxes marked with the radiate pattern, even though there was now no food in any of the boxes. Only a few bees visited the boxes marked with the other design.

In other experiments where I used geometric figures I had no success even though I continued the training for many days. The bees did not learn to distinguish a triangle from a square or a circle. I thought that they could learn to distinguish only the sort of figure which they normally met on flowers, and that perhaps they failed to learn the new patterns because they had never before encountered triangular or square colored patches (von Frisch, 1915). However Dr. Mathilde Hertz (1929a) discovered that bees could very easily learn to distinguish a triangle or a square, or any of the patterns in the upper row of Figure 19, from the rectangles or other shapes in the lower row of this figure. But they were incapable of learning to distinguish between the various patterns in the upper row, nor could they distinguish between two from the lower row.

From these and other results, Hertz (1929b, 1931) concluded that form-perception in bees and men is based on different criteria. For the eye of the bee the most important factor seems to be what Hertz has called the degree of "brokenness" of a pattern. All the patterns in the upper row of Figure 19 are clearly different for the human eye, which studies their form; but for the bees they are not distinct, because the brokenness of the figures is about the same. The length of the boundary, or line of contrast between figure and background, is small in these cases; all these figures are solid spots. But in the patterns of the lower row of Figure 19 the length of the boundary is considerable; these are extended figures. The bees seem to notice whether a figure is very much broken or is compact. Their form

Figure 19. Bees do not learn to distinguish between the shapes in the upper row of the figure or between the shapes in the lower row. But for the eye of a bee each shape in the upper row is distinctly different from every shape in the lower row.

perception is based in part on other principles than ours, a fact that may be related to the different optical arrangements of the compound eye of the insect on the one hand and the camera-like human eye on the other. But the chief reason for the difference may well be that bees see the patterns during flight. Since a bee's eye is rigidly fixed on her head, a broken pattern gives a flickering visual impression as the bee flies past.

In flying past she takes note of the changing brightness pattern. Her eye is well equipped to do this. Whereas its visual acuity is some one hundred times worse than ours, in a flickering light it can perceive as many as a ten-fold greater number of intensity changes per second than the human eye. A movie for bees would have to have a ten-fold more rapid sequence of frames in order to prevent flicker. With bees and other swift-flying insects the defective spatial resolving power of the eyes is compensated by a great power of temporal resolution (Autrum, 1949, 1952). According to these observations the form of the flowers probably contributes only to a modest extent to their discrimination by the bees. Their color is more important. Their most significant feature, their fragrance, will be discussed in the next chapter.

2. The Chemical Senses of Bees

In the first chapter I mentioned the fact that when a bee is visiting flowers it usually restricts its visits to a single species of plant. This is of advantage to the bee, since it encounters the same familiar mechanism within the blossoms. It is also very important for the plants, for it assures that the pollen brought by bees originates in other flowers of the same species. How is it possible for bees to recognize one species of flower among all the others that may be in bloom nearby? There are flowers of many colors and shades, but the number of different kinds of flowers is greater than the number of hues that the bee's eye discriminates. Maybe there is some other distinctive feature. Almost every kind of blossom has an odor which a man can tell from that of all other flowers. But we did not know, when I first became interested in such problems, whether the same is true for bees. We were not even certain whether bees could perceive the odors of flowers at all.

In studying this problem we may train bees to odors just as I had trained them to patterns, using experiments of the following kind. On an experimental table we place some cardboard boxes, each of which can be opened from above, and each provided with a hole in the front as an entrance for the bees (Figures 20 and 21). In one of the boxes we place a dish containing

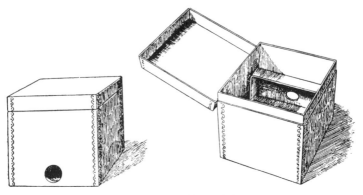

Figure 20. Left: Cardboard box for training to a scent.
Figure 21. Right: Cardboard box, opened. In training to the scent of an essential oil a small amount of this oil is dropped on the cardboard strip above the entrance to the box.

sugar-water and also a fragrant flower or a few drops of some essential oil. We change the position of this food box frequently to avoid training the bees to come to a certain place, for we wish only the odor to guide them to the food.

After several hours we take away all the boxes soiled by the bees and set out a new group of clean boxes.

Into one of them we drop a little of the scent that we have been using for training purposes, but now this box, like all the others, contains no food. The bees fly from one box to another, smelling around their openings. But they actually enter only the scented box. It is evident that they can smell this odor and that they use it as a guide to the source of food.

Next I wished to find out whether bees can distinguish as many different scents as we can. I therefore trained them to an essential oil made from the skins of Italian oranges. After one day of training I placed 24 boxes on the table (Figure 22), but only one contained

Figure 22. Arrangement for an experiment to learn whether bees are able to distinguish between several different scents.

the training scent, while the others were provided with 23 different essential oils. Afterwards I repeated the experiment with the same training scent and 23 new odors that had never previously been associated with food. In both experiments we counted the number of bees visiting the boxes during five minutes of observation. The box containing the training scent was entered by 205 bees in the first experiment and by 120 in the second.

Of all the other 46 boxes only three proved attractive

to the bees, and these contained the following three essential oils: "Essence de Cedrat" (148 visitors), "Essence de Bergamotte" (93 visitors), and an essential oil made from the skins of Spanish oranges (60 visitors). Only a few bees entered the remaining 43 boxes, or in some cases none at all.

These last three essential oils, which did attract a number of bees, were the only ones which were made from fruits of the same genus (*Citrus*) as the training scent. Their odor was very similar for the human nose and quite different from all the other scents. In a long series of further experiments it developed that the odor of the essential oil made from the skins of Italian oranges (the training scent) was noticeably preferred to the odors of Spanish oranges, Cedrat, or Bergamotte, but the difference in the numbers of bees attracted to these four scents was quite small.

From these and other experiments we may conclude that bees can distinguish between different qualities of odor just as well as a person whose sense of smell is very well developed. Furthermore, it seems clear that odors which are similar for the human nose are also similar for bees. In other experiments it turned out that substances without odor for us are likewise odorless for bees. The anatomy of the organs of olfaction is entirely different in bees and men, so it is surprising that their olfactory reactions are nevertheless so nearly the same (von Frisch, 1919).

Where are the sense organs of smell located in bees? Entomologists have known for a long time that insects can no longer react to scents if their antennae or "feelers" are cut off. The antennae were therefore believed by some to be the organs of olfaction. Other entomologists objected that such experiments were invalid because each antenna contains a relatively large nerve and because the cutting of this nerve would be a severe shock to the insect. The lack of response to odors might result from this shock and not from any loss of the sense of smell. There were conflicting views about this matter for many years. But it was possible to decide between them in the following way. Bees were trained to a certain odor, for example, peppermint oil. For these experiments cardboard boxes would have been cumbersome, since I wished to capture bees at the feeding dishes and operate upon them. I therefore used an arrangement like that described in the previous chapter

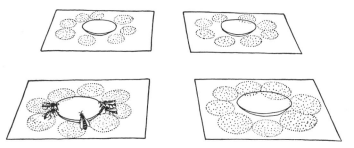

Figure 23. Bees are fed from a glass dish surrounded by a few drops of peppermint oil. The three other cards are provided with empty glass dishes and with another essential oil.

for training bees to color. The bees found the sugar solution in a small feeding dish placed in the open on a scented piece of cardboard. Close by were other pieces of cardboard with a different scent and without any food (Figure 23). After a short time the bees were able to distinguish the training scent from the other odor and seek out the correct card with certainty.

Next I caught trained bees just as they returned from the hive to the feeding place and cut off their antennae. When set free again, these bees continued to seek for the food but could no longer find it. Food was discovered only by chance—they were no more likely to visit the card with the training scent than any of the others. Such bees did not behave as though they were suffering from shock; they continued their search for food with vigor and persistence (Figure 24). But to be

Figure 24. The antennae have been cut off one of the bees that had been trained to the scent of peppermint oil. This bee is seeking for the training scent, but cannot find it.

certain, I trained other bees to find food on a blue card and then I removed their antennae. These bees found the blue card without hesitation and never confused it with a yellow one (Figure 25). This proved that the operation did not cause any severe shock, for the bees could still remember what they had learned but could not recognize a scent. In other words, the sense of smell is really located on the antennae.

A glimpse through the microscope shows that the

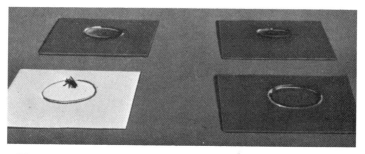

Figure 25. After being trained to the blue color a bee remembers what she has learned, in spite of the operation (cutting off antennae). She can distinguish the blue card (*lower left*) from the yellow ones, and alights on the empty glass dish seeking for food.

antennae of bees are densely covered by sense organs of varied appearance (Figure 27). These sense organs are not found on the long basal stalk, or scape, of the antenna or on the first three segments of its flexible, distal part, which is called the flagellum. They are confined to the eight outer or distal segments of the flagellum.

When I cut off the eight distal segments of one antenna (Figure 26,*bb*) and seven from the other (Figure 26,*aa*), the bee was still able to distinguish the training scent from other odors; it could even be trained to a new scent. But if the eighth distal segment of the second antenna was cut away the sense of smell was entirely lost. The experiment also shows that the effects of cutting away the antennae cannot be due solely to shock; the results can only be understood if we assume that the sense of smell is located on the antennae (von Frisch, 1921).

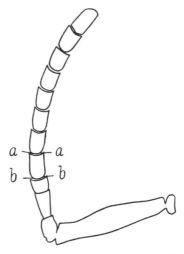

Figure 26. The antenna of a worker bee. After cutting along the line *aa* there is only one segment left that bears sense organs of smell. After cutting along the line *bb* no sense organs of smell remain on this antenna.

On the antennae seven different types of sense organs have been described. Which of these serve as organs of smell and what are the others for? Like all the rest of the insect's body, the antenna too bears a chitinous armor. This chitin would isolate the sensory cells inside the antenna completely from environmental stimuli were it not pierced by pores that are closed only by a delicate hair-like or disk-shaped chitinous structure and that indicate clearly the location of the individual sen-

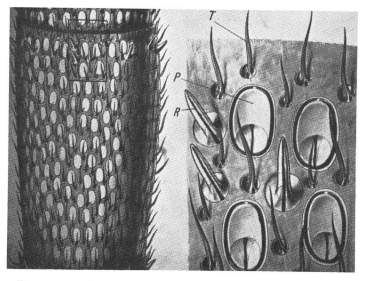

Figure 27. Left: One segment of the antenna at high magnification showing the sense organs. *Right:* The small portion outlined by black lines in the left-hand figure, greatly enlarged. *T*, organ of the sense of touch; *R*, cone-shaped organs; *P*, pore plates.

[43]

sory organs (Figures 27, 28, 29). There are certain sensory hairs with a fairly strong chitinous wall that are connected by a basal joint with the armored integument. Their construction alone reveals that they are tactile organs. The sensory cells end at the basal joint and are stimulated when the hair is moved at the joint by touching an object (Figure 28, where the chitin is shown in black).

Contact with the sensory cells of the organ of smell must be attainable by the olfactory substances. This may be possible with the other types of sensory cells, because their chitinous outer coat is extremely thin. For

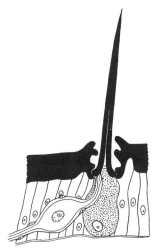

Figure 28. Section through an organ of the sense of touch. The nerve ending is situated at the base of the hair and is stimulated when the hair is moved.

various reasons it was to be suspected that the very numerous pore plates are organs of smell. Each pore is capped with an oval chitinous plate that is encircled by a thin furrow (Figures 27, 29), which electron micros-

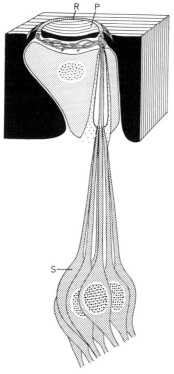

Figure 29. An olfactory sense organ (pore plate) on the bee's antenna. *S*, sensory cell; *P*, pore plate; *R*, circumferential furrow; chitin is black. Schematic; drawn by T. Hölldobler. (From von Frisch, *Aus dem Leben der Bienen*, 8th ed. [Springer-Verlag, 1969], p. 57.)

copy has shown to be perforated by numerous tiny openings. Here there may occur contact of olfactory substances with the sensory cells.

Yet are these pore plates actually organs of smell? If the olfactory sense could be abolished by destroying only the pore plates this assumption would be proved. But such an operation is impossible, because the pore plates are distributed over the antennal segments in contiguity with other sensory organs. Clarification was first afforded by the modern method of electrophysiology. As with the compound eye (p. 17) very fine electrodes can be inserted into the antenna and the changes of the electric potential ("excitatory potential") in the individual sensory cells can be measured. In this way it can be determined what sort of stimuli excite them. That sounds very simple. However, carrying out such experiments demands complex apparatus, a lot of time, and considerable skill. The results are unequivocal: the pore plates are olfactory organs. They are excited by floral fragrances and by other smells that are important to bees, for instance by the odor of their own scent organ (the "Nassanoff gland").

Up to now no other type of sense organs on the bees' antennae is known which could respond to odors. But among these there have been found types that respond to temperature, to the humidity of the air, or to carbon dioxide. All these factors are of importance for the atmosphere in the hive and are regulated by the bees.

There remain on the antennae a few types of sense organs of whose roles we are wholly ignorant even today (Lacher and Schneider, 1963; Lacher, 1964; von Frisch, 1967, pp. 497–499; Kaissling and Renner, 1968; Schneider, 1969). That the bees smell with their pore plates accords well with the fact that the number of pore plates on both antennae amounts to only about 6,000 in the case of a worker but to 30,000 in the case of a drone. A keen sense of smell is of basic importance to the male bee if he is to find a queen during the nuptial flight.

Most other insects also have the organs of smell confined to their antennae. But in several cases there are olfactory organs on other appendages of the head as well, i.e., on the palps of butterflies and of the water beetle *Dytiscus*.

I also tried to measure the sensitivity of the sense of smell in bees by training them to a certain odor and then diluting it more and more until they could no longer distinguish between scented and unscented boxes. When the dilution had reached a stage where the odor was no longer perceptible to us, we found that the bees too had lost the ability to perceive it. Thus the olfactory sensitivity of bees is roughly the same as that of human beings. We may be sure, therefore, that the scent of most flowers is insufficient to attract bees from a great distance.

The role of the sense of smell was well illustrated,

however, by experiments of the following type. I trained bees to distinguish a scented box with a blue front (Figure 30, *above*) placed among several others that were neither scented nor colored. Then I replaced this box with two others, on the right an uncolored box containing the scent and on the left an unscented box which was colored blue (Figure 30, *below*). The bees flew directly toward the blue box, but at a distance of about an inch they hesitated, apparently seeking for the

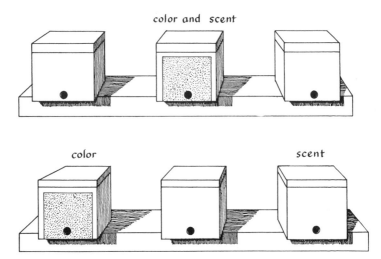

Figure 30. Bees are trained to a scented box with blue cardboard surrounding the hole (*above, middle box*). Afterwards, during the experiment, they find the blue color on the left and the scent on the right of the former feeding place (*below*). They fly toward the color from a distance of some meters, but they enter only the scented box.

scent to which they were accustomed. Then they inspected the other boxes, and when they approached the scented one they alighted and went inside. Apparently the color was visible from a greater distance, but the scent seemed to be the more convincing of the two. Likewise, the color of flowers has the advantage that it can attract bees from a greater distance, while the scent is specific for each species and thus permits the definite recognition of the flowers at close range (von Frisch, 1919).

There is one respect in which the sense of smell in bees is certainly superior to ours. The human organ of smell is located inside the nasal chamber, and the particles of odorous substances are brought to it by the stream of air utilized for breathing. Since this air stream is well mixed on its way into the nose, there can be no correlation between the shape of a scented object and the sensation of odor that it arouses. But in bees the olfactory organs are located on the antennae, and these can be moved about. Furthermore, the antennae are also covered by organs of the sense of touch, so that the sense of smell stands in a close relation to that of touch. A round scented object may give quite a different sensation to a bee than will an angular one. August Forel many years ago (1910) stated that bees might "smell" the form of objects as a result of this close relationship between the receptor organs of touch and smell on the antennae. The bees' association of touch and smell would be analogous to our own constant integration ever since infancy of what we see with our eyes and what we feel with our hands.

This association of touch and smell is very useful to bees as they visit flowers. Often they bring their antennae close to the flower, almost in contact with it, so that they can probably perceive even quite feeble odors. Moreover, if various parts of a flower have different scents, the bee can distinguish and locate these separate portions in a very precise manner. With this in mind I

Figure 31. The white flower of *Narcissus* with a yellow sap spot (stippled).

wondered whether there might be sap spots or nectar guides not only for the eye but also for the sense of smell. It was a pleasure to find that in many blossoms this was really true. In a narcissus, for instance, the yellow nectar guide (Figure 31) is not only of a different color from the white corolla but of a different scent as well. If we separate the yellow parts of the flower from the white parts and train some bees to one of these scents, they can distinguish between the two with certainty. And we too can perceive the difference in scent very easily once the parts have been separated, but we cannot do so by smelling the whole flower, because the two odors are mixed before reaching our olfactory organs. Bees, with their sense of smell localized on the surfaces of the antennae, can easily locate such differences and can be guided to the nectar by these scented sap spots. Such scented sap spots were also found in other flowers. Often there was no difference in quality but an increasing intensity of odor around the entrance to the bottom of the flower where the nectar is located (T. Lex, 1954; A. von Aufsess, 1960).

Human beings have two types of chemical sense, smell and taste. Anatomically, our sense organs of smell are located in the nasal chamber and those of taste on the tongue; they are supplied by different nerves. Physiologically, we can perceive through our sense of smell so many different qualities of odor that we even lack words to designate individual odors; but the sense of taste (when unaided by odors of foods) can distinguish only four qualities: sweet, bitter, sour, and salty. Moreover, the sense of smell is much more sensitive than that of taste. Consequently its biological significance is very different. Many animals with a well-developed sense of smell can perceive food or enemies at a considerable distance because evaporated particles can excite their olfactory receptors at very low concentrations. But chemical substances must be much more concentrated to excite the sense organs of taste, and since the latter are located in the mouth they serve for testing food when the animal is already in contact with it during a meal.

Much the same division of the chemical senses into two types is found in bees and other insects. The sense organs on the bees' antennae can detect, so far as we know, about as many different qualities of odor as the human nose. Since they are very sensitive they are able to detect food and other objects from some distance. But bees also have sense organs of taste located on the mouth parts in order to examine food when it is taken up. Bees can recognize a sweet solution only when they come in contact with it, but they are rather fastidious about sweetness. If offered a solution containing 20 per cent sucrose (about two-thirds molar), bees ordinarily suck it up. If it contains 10 per cent, we can see that there is an individual difference in taste—just as with men. Some bees drink, some hesitate, and others refuse it altogether. If a solution contains 5 per cent, they taste it but refuse to suck it up. This is their *threshold of acceptance*, and it varies depending on the feeding conditions. If there are many plants in bloom, one finds a high threshold of acceptance—sometimes about 40 per cent. During the fall when flowers become scarce there is a threshold of about 5 per cent. Bees, like men, become more fastidious under better conditions.

There is another limit, the *threshold of perception*, apparently the lowest concentration which will stimulate the sense of taste at all. This is invariable under good and poor conditions alike. To measure this threshold it is only necessary to starve the bees for several hours. Then they are ready to accept any solution that tastes noticeably sweet to them. For hungry bees the threshold is always between 1 and 2 per cent (von Frisch, 1934).

It is significant, biologically, that bees do not collect for storage in the hive solutions with a low concentration of sugar, although they may use them for their own nourishment. If stored in the honeycombs such solutions would not keep until winter. Indeed, all honey is chemically different from the nectar out of which it was made, and one important difference is an increase in sugar concentration by the elimination of water. Dilute sugar solutions require excessive periods of time, or excessive effort by the bees, in order to convert them into honey. Flowers have become adapted to this need by producing nectar with a high concentration of sugar (on the average about 40 per cent—R. Beutler, 1930; O. W. Park, 1928; A. Maurizio, 1960).

Chemists have learned that there are many different sugars, most of which taste sweet to human beings. But out of thirty-four sugars and sugarlike chemical compounds that we tested, only nine are sweet for bees: cane sugar, malt sugar, grape sugar, fructose, trehalose, melesitose, fucose, alpha-methylglucoside, and inositol. Most of the sugars which are sweet to us are tasteless to bees, for when offered either in pure condition or mixed with cane sugar, they have the same effect as pure water. This I found to be true for the following substances (those indicated with an asterisk are slightly repellent to bees): lactose, melibiose, cellobiose,* gentiobiose,* raffinose, tetraglycosan,* tetra laevan,* glycogen, galactose, mannose, sorbose, rhamnose, xylose, l-arabinose, d-arabinose, trimethylglucose,* beta-methylglucoside,* beta-methylfructoside,* beta-methylgalactoside, alpha-methylmannoside, erythritol, quercitol, mannitol, sorbitol, and dulcitol. It was not possible to determine definite rules valid for all insects governing the sweetness of these compounds, for substances which are sweet to one insect may be tasteless to others. For instance, the trisaccharide raffinose is taste-less to bees, but to ants it is the sweetest sugar of all those we have tested.

Honeybees are able to distinguish not only sweet but salty, sour, and bitter tastes as well. If we add salt to sugar-water the bees refuse it. I tried several methods to test the sensitivity of bees to salt. I could find the quantity of salt which had to be added to a sugar solution to cause its rejection, or, better, the amount which caused bees to hesitate slightly as they sucked up the solution. But a still better method is the following one, which depends upon the fact that bees take up larger volumes of solution if the sugar concentration is high. The sweeter the taste, the more they pump into their honey stomachs. For example, one of my experiments carried out on September 9, 1929, gave the following results, which are best presented in tabular form.

Number of bees	Concentration of sugar solution	Average volume of solution taken up per visit
70	17% (0.5 molar)	42 cubic millimeters
55	34% (1 molar)	55 cubic millimeters
49	68% (2 molar)	61 cubic millimeters

To find the threshold concentration of salt which the bees can barely taste, I offered them two dishes; both contained the same sugar solution, but one of the dishes also contained a very little salt. The amount of salt in the second dish was so low that the bees did not hesitate to take up the mixed solution, but they took less of it than of the pure sugar syrup. Apparently the pure solution tasted better. This sort of experiment enabled us to measure the very low concentrations at which they could taste the salt.

In another typical experiment performed on September 17, 1929, the bees were fed 0.5 molar sucrose, and again sucked up an average of 42 cubic millimeters per visit. Then, immediately afterwards, they were presented with a solution containing the same concentration of sugar plus 0.03 molar, or 0.017 per cent, salt (NaCl); from this mixed solution they took up only 37 cubic millimeters per visit. When the same test was repeated a year later the figures I obtained were 44 and 37 cubic millimeters, respectively. Even when a half-molar sucrose solution contains only 0.015 molar NaCl, the bees take up less than they take from pure half-molar sucrose. But the difference disappears if we use as little as 0.0075 molar NaCl.

I concluded from these experiments that honeybees are a little more sensitive to salt than are human beings. The same is true for materials which taste sour. But to bitter substances the bees are much less sensitive than we. They seem to enjoy a mixture of quinine and sugar which is so disgusting to the human sense of taste that anyone would spit it out at once (von Frisch, 1934).

With this in mind, I thought that I might be able to help the public authorities in the following way. In many countries beekeepers are allowed to buy sugar at low cost to feed their bees for the wintertime. Such encouragement to beekeeping serves to increase the number of colonies kept through the winter and thus improves honey production and the pollination of flowers and certain crop plants. But the cheap sugar thus placed at the disposal of beekeepers is supposed to be used only in the beehive, and not in the kitchen. To assure that this sugar was used only as intended, especially during the war years, it was desirable to mix something with the sugar which would make it useless for human consumption. But it was not easy to find a suitable denaturing substance. Either a compound was not efficient in producing a bad taste for men, or it had a disagreeable taste for bees as well. I proposed that the sugar be mixed with a small amount of octoacetylsucrose, which I had found to be very bitter for men but tasteless to bees. There was an initial difficulty in that this was a rare and expensive substance. But the chemical industry was able to work out a method for its production at low cost, and it was given the trade name "Octosan." It proved especially suitable for our purpose because it was a chemical compound of sucrose and acetic acid, and therefore quite harmless for both bees and men. Moreover, during the honey-ripening process in the beehive it decomposed after several weeks, so that its bitter taste was not imparted to the honey (O. Wahl, 1937).

Octosan was introduced in Slovakia, Bohemia, Poland, Romania, Bulgaria, Holland, and Belgium, and it proved practicable over a number of years. But in other countries the beekeepers would not agree to feed their precious bees such disgusting sugar. Perhaps some of them disliked this type of sugar chiefly because it was not useful for cooking.

It can be shown that salty, sour, and bitter are different qualities of taste for bees, just as they are for men. But the methods are somewhat complicated, because it is impossible to train bees to find food by taste as we trained them to find it by color or by odor. Let me cite only one example showing how we could determine that bitter and salty are different qualities of taste for bees. To a certain sugar solution we added just enough salt so that the bees rejected the mixture. Then we did the same for bitter, preparing a mixture of sugar and quinine just bitter enough to prevent the bees from accepting it. If these two substances had the same taste for bees, our two solutions should have aroused the same sensation of taste. However, we could show that this was not the case. For if we starved the bees, they took up a bitter mixture containing eight times more quinine than before. But these hungry bees showed no change in the threshold of refusal for mixtures of salt and sugar. Hungry bees will tolerate more bitterness, but they will not take a higher concentration of salt. Bitter and salty, therefore, cannot be the same quality of taste for bees. In a similar fashion it can be proved that sour is also a distinct quality of taste (von Frisch, 1934).

The fact that insects can distinguish four qualities of taste (sweet, bitter, salty, and sour) can also be demonstrated directly by experiments with water beetles (*Hydrous piceus, Dytiscus latissimus,* and *Cybister laterimarginalis*). These insects when seeking their food in the water can perceive the taste of food as it diffuses from some distance. Hence one can train them to find sources of food by the sense of taste. Water beetles trained to find food along with salty-, sour-, sweet-, or bitter-tasting substances can distinguish any one of these tastes with certainty from the other three qualities (A. Schaller, 1926; E. Ritter, 1936; L. Bauer, 1939).

Some animals have sense organs of taste not only in or near the mouth but on other parts as well. Professor D. E. Minnich of the University of Minnesota found (1922a) that butterflies could taste with the tips of their legs. If a butterfly steps into a sweet substance it instantly stretches out its proboscis, and often begins to feed. In this case the sense of taste is adapted to detect food as well as to regulate its intake. Therefore a lower threshold seems to be useful, for even a small amount of sugar can serve as a guide to a source of food. As a matter of fact, the taste receptors on the legs of butterflies are the most sensitive ones known to date. In flies, and other insects too, we find a well-developed sense of taste on the tips of the legs (D. E. Minnich, 1922b, 1929; F. Haslinger, 1935).

Figure 32. In fishes the sense of taste is not confined to the mouth. In this case (*Trigla*), certain rays of the pectoral fins are developed like fingers and bear many sense organs of taste.

Among the vertebrates we find low thresholds for the sense of taste only among the fishes. This fact is also related to the feeding habits of fish, for substances dissolved in water can guide a fish to food even from considerable distances. If the food is at some distance, materials diffusing from it reach the fish in a highly diluted state. Molecules from such food may reach any part of a fish's body, so we find that the sense of taste is not confined to the mouth, as in higher vertebrates, but is widely distributed over the surface of the body. *Trigla*, a bottom-dwelling fish, presents an analogy to the butterflies, for it is especially adapted for tasting materials lying at the bottom of the waters where it lives. Some of the rays of its pectoral fins are developed into fingerlike structures which bear many organs of taste (E. Scharrer, 1935). (See Figure 32.)

To illustrate the wide variation in sensitivity of the organs of taste in various animals, Figure 33 shows the

Figure 33. The bottle contains one quart of water. *Left:* The quantity of sugar (sucrose) which must be dissolved in this amount of water to give it a sweetness just perceptible (*a*) for bees (0.083 molar), (*b*) for human beings (0.0125 molar), (*c*) for a fish (0.0002 molar), and (*d*) for the legs of the red admiral butterfly (0.000078 molar).

amounts of crystalline sugar which must be dissolved in one quart of water to give it a degree of sweetness just perceptible to bees, to human beings, to a minnow, and to the legs of a butterfly.

Before concluding this chapter devoted to the chemical senses of bees I should like to describe some of the practical applications of this knowledge. Farmers often wish to cause more bees to fly to a certain kind of flower in order to improve its pollination. With red clover, in particular, the pollination is often very poor when fields are large. Under normal circumstances most of these flowers are visited and pollinated by bumblebees, which have a proboscis long enough to obtain all the nectar in the tubes of the clover flowers. Honeybees, therefore, often prefer other flowers which are in bloom at the same time and which are more convenient for them to feed upon. Bumblebees, on the other hand, are too scarce to effect the pollination of all the flowers in large fields of red clover.

About 1930, certain Russian scientists, knowing that it was possible to train bees to scents, developed a new method to improve the pollination of red clover. They brought beehives close to a field of red clover and trained the bees to the odor of these flowers. The training was accomplished during the evening by feeding the bees in the hive sugar solutions scented with the flowers of red clover. To obtain this scented sugar solution the Russians simply put some flowers into the sugar-water so that after several hours the odor of the clover was taken up by the solution. The flowers themselves were then strained off. Many of the bees which had been fed in this way flew to red clover plants the next morning. By using this technique the Russians managed to improve the crop of red clover to a considerable degree.

In 1927, stimulated by a beekeeper's observations, I had proposed a similar method; to explain it I must anticipate a little of the next chapter. If bees have discovered a good feeding place they announce the fact in the hive by means of certain dances performed on the honeycombs. The other bees not only learn that there is food available, but they are also informed in which flowers it is to be found. They obtain this information from the scent of the flowers which adheres to the bodies of the dancing bees. If we feed some bees at an artificial feeding place provided with the scent of the flowers that we wish them to visit, the foraging bees will perform their dances in the hive and stimulate other bees, which will fly out in search of the same odor and thus reach flowers of the same species. On this basis I expected to guide many bees to a certain field in order to secure better pollination. Because of the catastrophic shortage of food in Europe in consequence of the Second World War, I studied this practical problem in cooperation with beekeepers and agricultural experiment stations. Out of several methods which gave good results let me describe only a simple one.

When a field of red clover came into bloom we brought several beehives close to the field and fed the bees every morning with a small amount of sugar-water in a box placed just in front of the hive (Figure 34).

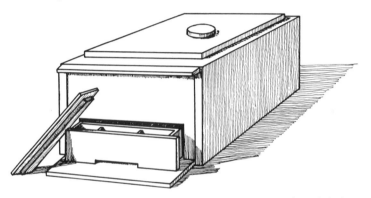

Figure 34. Beehive with feeding box, the cover of which is opened to show the three compartments inside (see Figure 35).

The box was divided into three compartments; the middle one contained flowers of red clover and the others contained the food (Figure 35). To reach the sugar the bees were obliged to creep through these flowers, and thus their bodies were certain to become heavily scented. Afterwards they danced in the hive, and other bees, perceiving from the bodies of the dancers the odor of the flowers in our feeding box, searched for red clover when they flew out to forage.

To test the effectiveness of this method we performed experiments involving two fields of red clover so far apart that the bees visiting one field could not possibly fly to the other. The quality of the soil, the size of the field, the seed, the fertilizer, and other such factors were kept as similar as possible. Near each field we placed the same number of beehives. At one field the bees were fed every morning with sugar-water pro-

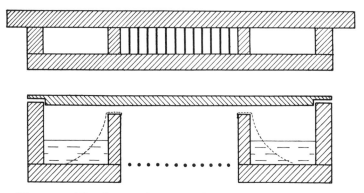

Figure 35. The feeding box, seen from above (*upper drawing*) and in cross section (*below*). The box is set in front of the entrance of the hive, so that the bees can enter and leave the hive by passing under the middle compartment of the box. The middle compartment must be filled with the flowers that one wishes to be visited by the bees. The lateral compartments contain sugar-water. To reach the sugar the bees pass the wire grid and crawl through the flowers. Afterwards, when dancing in the hive, they carry the scent of these flowers on their bodies.

vided with the scent of red clover as described above. The bees at the other field received the same amount of sugar-water, at the same time of day, but without any scent; this field served as our control. We repeated this experiment twelve times in various localities. In all cases the field where the bees were guided by odor was visited by more bees than the control field—on the average, three to four times as many. The yield could be measured in nine of the twelve experiments, and the weight of the crop averaged 40 per cent higher at the fields where scented sugar-water had been employed (von Frisch, 1947). Later experiments confirmed the success of this method (von Rhein, 1957).

Procedures of this kind can be useful to beekeepers as well as to farmers. In many experiments we found that when bees were directed to seek for certain flowers the yield of honey could be increased by 50 per cent or more if they were guided in this way to flowers which were supplying a rich source of nectar. For bees which have been fed scented sugar-water seek very diligently for the flowers whose scent was used at the feeding place, and they collect nectar more industriously than bees which have been fed unscented sugar (von Frisch, 1947).

In the actual employment of these methods one must pay careful attention to some factors which I cannot discuss here in detail. Unless these are known, one will not always be successful. That is probably why this method of "guiding by odor" has been adopted in only a few countries. For many years it has been widespread in the USSR, where the combination of agricultural and beekeeping activities in collectives facilitates the uniform application of such measures. When some day in the future food grows scarce, people in other countries too should recall that in their own language bees can be aroused to greater industry and can be dispatched to definite species of flowers in accord with the wishes of the beekeeper and the farmer. I have mentioned it here to present a new example of the well-known fact that research work performed for its scientific interest alone often proves later to be of great practical value in ways which could never have been foreseen.

3. The Language of Bees

When I wish to attract some bees for training experiments I usually place upon a small table several sheets of paper which have been smeared with honey. Then I am often obliged to wait for many hours, sometimes for several days, until finally a bee discovers the feeding place. But as soon as one bee has found the honey many more will appear within a short time—perhaps as many as several hundred. They have all come from the same hive as the first forager; evidently this bee must have announced its discovery at home.

I was curious to learn how bees could tell their fellows about the presence of food at a new location. But it is not possible to observe what happens as the bees crawl about between the honeycombs inside an ordinary beehive. I therefore constructed an observation hive in which the honeycombs were arranged edge to edge so that they formed one large comb, the surface of which could be watched through glass windows (Figure 36). It was also necessary to number every bee we wished to study, so that it could be recognized individually among the mass of other bees on the honeycombs. For this purpose I painted the bees with small spots of five different colors; a white spot on the front part of the thorax stood for the number 1; a red, for 2; a blue, for 3; a yellow, for 4; and a green, for 5. A white spot on the hind part of the thorax indicated the number 6; a red, 7; a blue, 8; a yellow, 9; and a green, o. By employing combinations of these colors I could

Figure 36. The observation hive. The entrance for the bees is at the right.

apply any two-digit number; and spots to indicate the hundreds were painted on the abdomen, so that I could number as many as 599 bees at one time (Figure 37). The actual colors applied to the bees consisted of dry artist's pigment mixed with a solution of shellac in alcohol. This mixture dries quickly and adheres well to the body of a bee.

To study the behavior of bees which have just discovered a rich source of food one may set out near the observation hive a glass dish filled with sugar-water. When a foraging worker comes to this feeding place she is marked with a colored spot while she is sucking

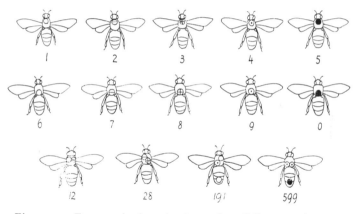

Figure 37. For numbering the bees, five different colors were used. The meanings of the symbols in the drawing are: white circle, a white spot; horizontal line in the circle, a red spot; cross in the circle, a blue spot; dot in the circle, a yellow spot; black circle, a green spot.

up the sugar, so that we can recognize her later in the hive. After she has returned to the hive our marked bee is first seen to deliver most of the sugar-water to other bees. Then she begins to perform what I have called a round dance. On the same spot she turns around, once to the right, once to the left, repeating these circles again and again with great vigor. Often the dance is continued for half a minute or longer at the same spot. Frequently the dancer then moves to another spot on the honeycomb and repeats the round dance and afterwards ordinarily returns to the feeding place to gather more sugar.

During the dance the bees near the dancer become greatly excited; they troop behind her as she circles, keeping their antennae close to her body (Figure 48). Suddenly one of them turns away and leaves the hive. Others do likewise, and soon some of these bees appear at the feeding place. After they have returned home they also dance, and the more bees there are dancing in the hive the more appear at the feeding place. It is clear that the dance inside the hive reports the existence of food. But it is not clear how the bees that have been aroused by the dance manage to find the feeding place.

To learn whether the round dance imparts information about the direction in which food is to be found, I fed several numbered bees from my observation hive at a feeding place 10 meters to the west. At each of four points in the meadow around the hive, to the north, south, east, and west, I placed on the ground a glass dish containing sugar-water scented by a little honey. A few minutes after the start of round dances in the hive new bees appeared simultaneously at all the dishes regardless of their direction. The message brought by a bee as she performed the round dance seemed to be a very simple one, one that carried the meaning "Fly out and seek in the neighborhood of the hive."

Figure 38. Numbered bees are fed on cyclamen flowers.

But it is not natural for bees to gather their food from glass dishes. If we feed our numbered bees at the same feeding place with sugar solutions placed upon fresh flowers, for instance, cyclamen (Figure 38), and if the foraging bees dance after they have returned to the hive, new bees fly out as before, but now they are seeking for something definite. Somewhere in the vicin-

Figure 39. The newcomers sent out by the dancing bees coming from cyclamen are interested only in this species of flower. They alight on cyclamen and seek for food there. They are not attracted to phlox or other flowers.

ity of the hive we establish an observation point by placing on the ground two large dishes; one of these contains cyclamen flowers and the other contains phlox. The newly aroused bees are interested only in the cyclamen; they take no notice of the phlox (Figure 39). Next we can change the flowers at the feeding place, putting food on blossoms of phlox. Now the same numbered bees that had previously collected sugar-water on cyclamen flowers begin to gather it on the phlox (Figure 40). After a few minutes the situation at the observation point changes radically: the new bees are no longer interested in cyclamen; they alight only on phlox—examining the flowers as though convinced that they must contain food.

Figure 40. Feeding on phlox.

I succeeded in obtaining similar results from this type of experiment whenever I employed fragrant flowers, even those with a very feeble scent; but I did not succeed when I chose flowers without any odor at all. If, for instance, I fed numbered bees on the unscented blossoms of bilberry (*Vaccinium myrtillus* L.), then the new bees that swarmed out of the hive searched earnestly in the vicinity for the food that had been announced to them; but a dish of bilberry placed in the meadow did not receive any more attention than the surrounding grass or other unscented objects. The same result was obtained in other experiments with odorless blossoms (grass blossoms, *Holcus lanatus* L., the lily *Tritonia crocosmaeflora* Voss., and the Virginia creeper *Parthenocissus quinquefolia* Planch.).

On the other hand, the same result can be obtained without employing flowers at all. We can feed several numbered bees from glass dishes of sugar-water, each dish resting upon a piece of cardboard scented by peppermint oil. Nearby we also set out a series of similar cards, some scented with a few drops of peppermint oil and some with other essential oils. The new foragers are now interested only in the scent of peppermint; they alight on every place and on every object touched by this oil. Apparently the newly aroused bees learn the scent of the flowers just visited by the dancer, and when they fly out seeking for this odor they reach the same kind of flower. Here is a biological function of flower scents which had not been known before.

The new foragers remember very well the odor that they learned from the dancer, and they are able to find it with certainty. This was once demonstrated to me in striking fashion in Munich during mid-July at the systematic section of the botanical garden, where I counted seven hundred different plants blooming at the same time. One of these was *Helichrysum lanatum* DC., planted only in one small flower bed and visited only by one species of solitary bee, but never by honeybees. Biologists have reported that honeybees have never been seen gathering nectar from this particular species. But I fed some numbered bees from my observation hive at the border of the systematic section of the botanical garden, supplying sugar-water in a glass dish surrounded by several blossoms of *Helichrysum*. The dancing bees coming from this feeding place carried the odor of *Helichrysum* on their bodies. Within the following hour many honeybees visited the flower bed containing *Helichrysum lanatum* and alighted on the flowers in search of food. They had sought out the specific odor of this species among the seven hundred other scented flowers blooming nearby (von Frisch, 1923).

The bees that troop after the returning forager as she dances on the honeycomb perceive the flower scent in two ways. By holding their antennae toward the dancer they smell the scent adhering to her body as a result of her contact with the flower. The upper surface of the bee's body has the ability to hold scents for long periods (Steinhoff, 1948). Second, during pauses in the dance, the dancer feeds the bees that are following her by regurgitating a droplet of nectar from her honey stomach. This nectar was gathered from the bottom of the flower and is saturated with its characteristic scent. The bees that have been aroused by the dancer can thus receive the odor of the flower both from her body and from the nectar that she passes to them.

The reader may well ask how we can know these details. It was actually not difficult at all to find them out by means of experiments of the following sort. We can feed unscented sugar-water to several numbered bees as they alight on cyclamen flowers. To prevent this sugar solution from taking up the odor of the surrounding cyclamen blossoms we place it in a small, spherical glass flask from which the bees can suck it up only through a narrow cleft (Figure 41, *below*). These bees carry the odor of cyclamen flowers only externally on their bodies, and the other bees which they send out from the hive seek for this odor. In another experiment we allow different foraging bees to alight on an odorless card and feed them in the same manner with sugar-water that we have previously exposed for one hour to the odor of cyclamen flowers (Figure 41, *above*). In this case the odor is carried only internally, in the honey stomach, but the bees that are aroused by the dancers again seek for cyclamen.

It is interesting to determine the relative effectiveness of these two methods of transporting the flower scent back to other bees in the hive. We can put the matter to a test by feeding bees as they alight on cyclamen flowers with sugar-water that has been saturated with the odor of phlox blossoms. Now the scented material adhering to the body of the dancing bee will have a dif-

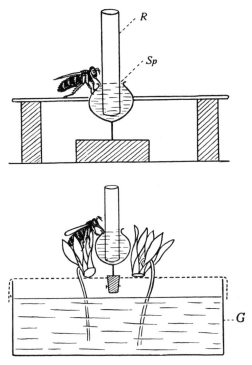

Figure 41. Below: The bee sitting on cyclamen takes up odorless sugar-water. She carries the odor only externally on the body. *G*, glass vessel containing water. *Above:* The bee sitting on a piece of unscented cardboard takes up scented sugar-water. She carries the odor only internally in the honey stomach. *R*, glass tube; *Sp*, the narrow cleft from which the food can be taken.

ferent odor from the nectar carried in her honey stomach. If the feeding place is close to the hive, the bees aroused by the dancer visit cyclamen flowers about as often as phlox. But if the feeding place is farther away—for example, a half mile from the hive—the new foragers are interested mostly in phlox, the odor of which was carried in the honey stomach of the dancer. This means that during a long flight back to the hive the scented material adhering to the body of a bee is decreased. Clearly the scent taken up with the nectar and carried in the closed honey stomach (Figure 42) is very important when bees are collecting at considerable distances (von Frisch, 1946a; 1967, pp. 224–227).

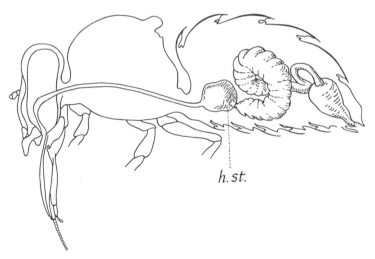

h. st.

Figure 42. The intestinal tract of a bee. *h. st.*, honey stomach.

These dances are observed only if there is a rich source of food (Figure 43), so that evidently they also carry the basic meaning "There is plenty of food and sweetness." If we take away a glass dish full of sugar solution from which bees have been feeding, and replace it by another dish provided merely with some sheets of filter paper moistened from beneath by a little

Figure 43. Plenty of food.

sugar-water, then there is a scarcity of food although the composition of the solution is unchanged (Figure 44). Provided that we do not reduce the volume of sugar-water below a certain critical level, the bees will continue to collect it by sucking it up laboriously from the pores of the filter paper. But now they do not dance after returning to the hive, and hence no new worker bees appear at the feeding place. The same result follows if we dilute the sugar solution to a certain point, even though a large volume of the fluid is still available. The sweeter the sugar, the more vigorous are the dances (von Frisch, 1923, 1934).

Figure 44. A scarcity of food. Filter paper moistened from beneath with sugar-water by the syringe.

This reaction can also be observed under natural conditions when the bees are visiting flowers; and it is important for the bees, since several kinds of flowers often come into bloom at the same time. When this happens the various species may be discovered by foragers from the same hive, but those bees which have discovered the richest source of food dance most vigorously and send out the largest number of new foragers to this kind of plant. The pollination of these plants is thereby guaranteed in the best manner, and likewise the beehive has the benefit of securing the best and sweetest nectar. After the bees have harvested a rich source of food so industriously that the nectar of this species becomes scarce, the dances come to a stop and no more worker bees are sent out to visit this kind of flower. In this way the activity of the whole group of foraging bees is adjusted to the relative abundance of the nectar of various flower species.

Figure 45. Three bees at the glass dish with plenty of food (sugar-water). The opened scent organ ("Nassanoff gland") can be seen in the bee at the left beneath the arrow. The bee at the right has already withdrawn its scent organ.

The observations that I have just described make it clear that the dancer guides other bees to the flowers she has discovered, and uses the flower odor as an indication of the location of the rich food source. Another scent is also used for this purpose, and it is produced by the worker bee herself. Bees have a scent organ located on the abdomen in a pocket of skin lined with glands. Usually the pocket is closed and cannot give out any scent. But bees returning to a rich source of food open the scent organ as they approach the feeding place and alight upon it (Figure 45). In doing this they apparently apply to the food source a scent which is very attractive to other bees. It seems to carry the meaning "Come here; this way!" The importance of this scent when bees are visiting flowers which have no odor could be shown by the following experiment: I arranged two feeding places without scent, and at each place allowed about twelve numbered bees to collect sugar-water. At one feeding place I applied a little shellac to each bee as she arrived, so as to close the pocket containing the scent glands. These bees continued to dance in the hive just as vigorously as before, but they could not give out any scent when they returned to the feeding place. As a result I counted at this dish only one-tenth as many newcomers as at the control dish, where the bees had been allowed to use their scent organs in the normal way (von Frisch, 1923).

For many years in performing experiments of this general kind I always placed the food in the immediate vicinity of the hive, partly for convenience and partly so that I could watch the bees at the same time both at the observation hive and at the feeding place. Occasional observations suggested that bees could also tell something about the distance from the hive to the feeding place. Since bees often gather food a mile or more from the hive, it would clearly be advantageous if a forager which had located a rich but distant source of food could convey to other bees some idea of its location as well as its odor. To study this interesting question I performed the following experiment in August, 1944. Two feeding places were arranged, one 10 meters and the other 300 meters from the hive; both were visited by numbered bees from my observation hive. In order to

Figure 46. A small wooden table is used as feeding place, and it can be fixed in the ground anywhere by means of its single leg, which is pointed at the lower end. On the table is a glass dish with sugar-water, resting on a scented card. The photograph shows many bees visiting the feeding place. Usually we restricted the numbered bees to about ten for each experiment.

learn from experiments of this sort where the newly aroused foragers may actually fly in search of food, we used a scent such as lavender oil in the following way. At one of the feeding places sugar-water was offered in a glass vessel, and this vessel rested on a card scented with lavender oil (Figure 46). At different points in the

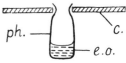

Figure 47. Left: The bees sent out by the dancers smell around the three tiny vials containing the same odor as that supplied at the feeding place, and they alight on the cardboard. *Right:* The drawing shows one vial containing the essential oil, set in a hole of the cardboard. *ph.*, vial; *c.*, cardboard; *e.o.*, essential oil.

meadow we also set out test cards that were without food but were supplied with small vials of the same oil. The bees aroused by dancers from the feeding place sought for this scent, and on approaching one of the test cards they would often alight upon it (Figure 47).

After this preparation I was ready to begin the actual

[86]

experiment by giving plenty of food at the nearer feeding place, so that bees returned from a rich source at 10 meters and danced in the hive. After several minutes there were many newcomers at this feeding place and on other test cards close to it, but only a few were seen near the 300-meter site. For example, in one experiment —and by no means the most striking one—we counted 12 new bees near the feeding place 300 meters from the hive and 174 during the same time at the 10-meter site. But when we reversed the procedure, giving plenty of food at the distant source and none at 10 meters, then the only bees dancing in the hive were those fed at 300 meters. Now the newcomers appeared quickly at the distant site, while we saw only a very few at 10 meters from the hive. During one hour of observation in a typical experiment of the latter type, we counted 61 bees near the distant feeding place and only 8 near the feeding place 10 meters away. Often the difference was even greater. Hence it seemed clear that the returning foragers brought a message about the distance of the rich food source they had found.

In order to study the nature of this message we can give plenty of food at both feeding places at the same time and mark the bees collecting food at each. When we now look into the observation hive we see a truly curious sight: all the bees marked at the 10-meter food source are performing round dances just like those described above (Figure 48). But all the bees that have come from the more distant feeding place are dancing in quite a different manner. They perform what I have called a "tail-wagging dance." They run a short distance in a straight line while wagging the abdomen very rapidly from side to side; then they make a complete 360-degree turn to the left, run straight ahead once more, turn to the right, and repeat this pattern over and over again (Figure 49). This wagging dance was one that I had observed many years before; but I had always

Figure 48. The round dance. Three bees following the dancer are receiving information.

[88]

taken it for the characteristic dance of bees bringing pollen to the hive, whereas now I saw that it was performed most vigorously by bees which were bringing in sugar solutions from the experimental feeding place at 300 meters.

It soon became clear that my original interpretation had been incorrect. The error arose because I had always

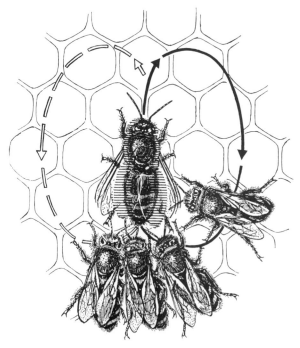

Figure 49. The tail-wagging dance. Four followers are receiving information. (From von Frisch, *Tanzsprache und Orientierung der Bienen* [Springer-Verlag, 1965], pp. 29 and 57.)

furnished sugar-water at feeding sites close to the hive, chiefly for convenience of observation. The pollen carriers, on the other hand, were arriving from their natural feeding places on flowers some distance away. (Bees carrying pollen can easily be distinguished in the observation hive by the tightly packed pollen baskets formed by stiff hairs attached to the hind legs.) Hence I had seen round dances performed only by bees which were gathering nectar or sugar-water, and wagging dances only by pollen carriers. But once I realized that the two types of dances were related to the *distance* of the food source and not to the nature of the food, it was easy to show by suitable experiments that the dances of pollen gatherers are no different from those performed by nectar gatherers returning from the same distance.

It seems as though a happy dispensation from my scientific guiding star allowed me to discover this error myself. But let younger investigators be warned by this example, as they strive impatiently to publish their results after long years of frustration. Let them test their findings doubly and trebly before they regard any interpretation as certain. For often nature reaches her goal by another path, where man cannot see his way.

I also modified the experiments described above by gradually moving the nearer feeding place from 10 meters to greater and greater distances; as I did this, the same group of numbered bees continued to return to the dish. But between 50 and 100 meters the round dances gave way to wagging dances. At the same time I moved the more distant feeding vessels nearer and nearer to the hive; and as I did so, the same group of bees continued to collect from it. Between 100 meters and 50 meters the wagging dances of this group changed to round dances. It was clear that the round dance and the wagging dance are two different terms in the language of bees, the former meaning a source of food near the hive and the latter a source at 100 meters or more (von Frisch, 1946a,b).

The dances are apparently understood by the bees in the hive, as could be shown by the following experiment. One group of numbered bees was induced to gather food at a feeding place 10 meters from the hive, while another group was fed at a second site 200 meters away. At both locations the feeding vessels were situated on unscented cards. We then stopped supplying sugar at both places, and allowed the dishes to remain empty for an hour or two. After this time most of the bees from both groups were sitting inactive in the hive; only from time to time would one of them fly out to the feeding place to see if anything was to be had. If we now refilled the dish at the more distant site, then the wagging dances of the first gatherers to return with full stomachs aroused chiefly bees from the group which had previously visited the distant feeding place. But when we offered sugar-water at the nearer site, then the resulting round dances aroused mostly bees which had previously been feeding there.

When bees are actively collecting food from flowers they often fly out to a distance of a mile or more from their hive. Under these circumstances a message telling only that a food source was nearer or farther than 100 meters could scarcely have much biological value. As a matter of fact, the wagging dance not only announces

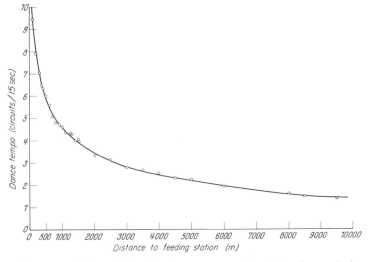

Figure 50. The relationship between tempo of dancing and the distance to the feeding place. Average values from the experiments of Bräuninger, von Frisch, Heran, Knaffl, and Lindauer. The curve is based on 6,267 measured dances (see von Frisch, 1967, p. 69, Table 6). (From von Frisch, *Tanzsprache und Orientierung der Bienen* [Springer-Verlag, 1965], p. 70; *Dance Language and Orientation of Bees* [Harvard University Press, 1967], p. 69.)

that there is a rich source that is far away; it also tells *how* far away. We can determine the distance quite accurately from the number of turns in the wagging dance that are made in a given time. By gradually moving a feeding site to greater and greater distances we reached 10 kilometers (6.2 miles), after rather strenuous experiments. We counted the number of turns per unit time made by dancers returning from distances running from 100 meters up to 10 kilometers. Figure 50 shows the result of 6,267 such observations. At 100 meters there were 9 to 10 complete cycles of the dance within 15 seconds; for 300 meters, about 7; for 500 meters, 6; for 1,000 meters, 4.6; for 2,000 meters, 3.3; for 5,000 meters, 2.2; for 10,000 meters, 1.25 (Knaffl, 1953). At very great distances the wagging dance often consists only of the long linear wagging run and is then discontinued. This circumstance and other observations have convinced us that the linear run in the dance figure shows the bees the distance, indeed by means of its duration, which is strongly emphasized and sharply delimited through the dancer's waggling motions and by a simultaneous buzzing sound. According to our measurements the "duration of waggling" lasts about ½ second for a distance of 300 meters, about 1 second for 500 meters, 1 ⅓ second for 1,000 meters, 2 seconds for 2,000 meters, 4 seconds for 4,500 meters—it grows with increasing distance to the goal. The sound during the waggling movements consists of a series of short bursts of tone (Figure 51) that are produced by the flight muscles in the thorax. In one second there is a series of about 30 bursts; being about 250 Hertz (cycles per second), the frequency of vibration of these short individual tones conforms with the wingbeat frequency (Esch, 1961, 1964; Wenner, 1962). The bees that are in contact with the dancer (Figure 49) can perceive the sound and the waggling movements with their delicate sense of touch and thus are informed how far they need to fly.

In order to point out the distance correctly, the bee has to measure it during her flight. It is clear that this is not done with a ruler. The bee also does not indicate the true distance of the goal in meters. When in her flight to the feeding place she has a head wind, her dances indicate a greater distance and with a tail wind a shorter distance, than in a calm. Flight up a steep slope also occasions the indication of a greater distance. Her standard is the expenditure of energy for the distance she covers. The reader will ask how we know that. With an auto the distance traveled can be measured by the amount of fuel used. Going uphill more is needed. The bee's fuel is the sugar that she has in her blood. This sugar content of the blood can be determined precisely. Experimentally it has been possible to show that their indications of distance are dependent on the consumption of sugar (Scholze, Pichler, and Heran, 1964).

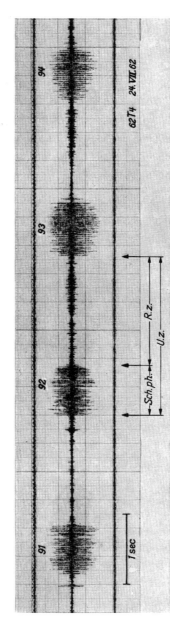

Figure 51. Sample recording of sound production during three cycles of a tail-wagging dance. Feeding station 300 meters distant. *Sch. ph.*, tail-wagging phase; *R.z.*, duration of return run; *U.z.*, duration of total circuit. (From H. Esch, *Zeitschrift für vergleichende Physiologie,* vol. 48 [1964] p. 535, Fig. 1.)

When the feeding place is near the hive, bees which have been aroused by the round dances fly out in all directions and seek for food in the immediate vicinity. If the source is farther away, the bees learn from the wagging dance, as we have seen, the distance at which the newly discovered food lies. But in addition they learn the *direction* in which they must fly. This surprising fact can be demonstrated by experiments of the following type: We feed some numbered bees on a scented card situated 200 meters south of the hive. Other cards, bearing the same scent but without food, are set out on all sides of the beehive at a distance of 200 meters. Within a few minutes many new bees appear, not only at the feeding place but at the other cards lying nearby. But at the cards lying in other directions we do not see a single bee, or at most a very few. If we shift the food to one of the other directions, say, to the card 200 meters *east* of the hive, we find after a short time that newly aroused bees fly out to the east. The language of bees is truly perfect, and their method of indicating the direction of food sources is one of the most remarkable accomplishments of their complex social organization.

If we observe dancers which have returned from a feeding place whose location is known to us, it is surprising to see that all these bees perform the same dance; in particular, they always head in the same direction during the straight part of the wagging dance (see Figure

49). In a typical case the bees collecting at a food source 200 meters south of the hive danced on the honeycombs in such a way that the straight portion of their dance was always headed to the left. If at the same time other bees were gathering sugar-water from a feeding place 200 meters north of the hive, we saw that they pointed to the right during the straight phase of the dance. In other words, the direction of the straight part of the wagging dance is related in some way to the direction of the food source.

When we watched the dances over a period of several hours, always supplying sugar at the same feeding place, we saw that the direction of the straight part of the dances was not constant, but gradually shifted so that it was always quite different in the afternoon from what it had been in the morning. More detailed observations showed that the direction of the dances always changed by approximately the same angle as the apparent motion of the sun across the sky. This was not entirely unexpected, since experiments carried out several years before by Wolf, Santschi, and others had shown that both bees and ants often use the sun as a sort of compass when traveling over a level plain without conspicuous landmarks. These earlier observations had involved bees and ants which went out from their homes in search of food and then returned before the direction of the sun had changed appreciably. If a worker bee has flown toward

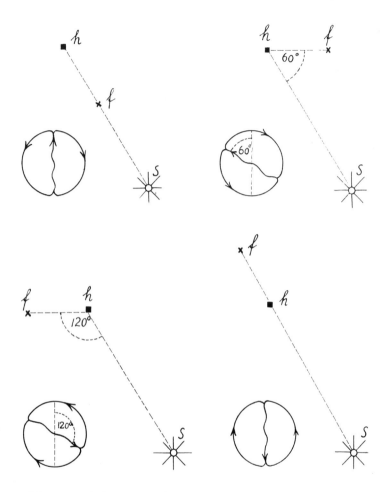

Figure 52. *h*, beehive; *f*, feeding place; *s*, sun's position. At left of each diagram is shown how the bees dance on the perpendicular comb to indicate the direction of the feeding place with respect to the sun's position. Note that the bearing of the sun is transferred to the upward direction, perceived by means of gravity.

the sun in leaving the hive, she has only to fly directly away from the sun to find her way home. If the sun is to her right on the trip away from the hive, the bee keeps it on her left during the flight home. Hence, when we found that the straight part of the wagging dance shifted with the sun's position, it became clear that these dances also indicated the direction of the feeding place with reference to the sun.

The key to an understanding of this message is a very curious one. We must recall that under normal circumstances a bee dances on the perpendicular honeycomb inside a hive where it is quite dark. In the ordinary hive bees cannot perceive the direction of the sun, but apparently rely instead on the direction of gravity. They orient the straight portion of the dance at the same angle to the force of gravity as the angle they have flown with respect to the sun during the flight from hive to feeding place (Figure 52). If a dancer heads directly upward during the straight part of her dance on the honeycomb, this apparently means "The feeding place is in the same direction as the sun." If the straight run points down, it means "Fly away from the sun to reach the food." If during the straight portion of the dance the bee heads 60 degrees to the left of vertical, then the feeding place is situated 60 degrees to the left of the sun. Similarly, a dance with its straight run pointed 120 degrees to the right of vertical indicates a feeding place situated 120 degrees to the right of the sun's position in the sky (von Frisch, 1946a,b, 1948).

It is remarkable that the heading *toward the sun* is the direction of flight selected to correspond with an *upward* movement during the straight component of the wagging dance. One cannot believe that bees decided all at once to arrange matters thus. We may be sure that this meaningful relationship has developed gradually, like other abilities, in the course of the history of the species. I will return to this question at the end of this discussion.

Even with an overcast sky the dancers point out the direction with reference to the position of the sun. At all events they cannot do so when the sun is darkened by heavy rain clouds. But when the weather is like this the bees don't go flying out and need no direction signs. When the sky is covered by a cloud layer that is not altogether too dense, but that nonetheless renders the sun completely invisible to us, then the dances are not deranged. That is noteworthy and was the cause of many experiments. The thought nearest to hand was that the bees know where the sun is at a given time, even when it is invisible just then. They have an excellent memory for time. When they are fed at a certain place from 11–12 o'clock, on the next day at the same hour they go hunting about there for food, even if there is none there. From other experiments we know that they are also familiar with the diurnal course of the sun (Beling, 1929; Renner, 1957, 1959; von Frisch, 1967, pp. 347–356). Perhaps they remember, even without seeing the sun, where in the familiar landscape it is at a given time. The following experiment showed that that is not the right explanation: on a day with a completely overcast sky we moved the hive into a distant area with which the bees were unacquainted. There we set up a new feeding place. It was impossible that the foragers should know where the sun is in that neighborhood at every hour of the day. Despite this they danced perfectly. They must have made the sun out through the cloud cover.

Thereupon I studied the bees' dancing when the sky was overcast while they were viewing it through colored filters that transmitted only a certain narrowly limited spectral range. It turned out that they could perceive the sun behind the clouds only beneath those filters that transmitted the ultraviolet rays, which we cannot see. Beneath a glass plate that was fully transparent for us but that absorbed the ultraviolet (400–300 mμ), their dances were disoriented. But under a glass plate that was black and opaque for us, while transmitting the ultraviolet, they danced perfectly correctly. That was surprising, for any physicist will object that the rays of long wave length penetrate a cloud cover better than those of short wave length. But with sensitive photographic plates it was always possible to show in the ultraviolet region a brightening at the sun's position, when the cloud cover was such that the bees could still perceive the sun. That it is perceived by the bees is to be ascribed to the particularly great sensitivity of their eyes for ultraviolet (see von Frisch, 1967, pp. 366–377).

The bees which are aroused by the dances on the honeycombs recognize the angle of the dance relative to gravity, and in flying out to the food they remember this angle and relate it to the position of the sun. We did many experiments in order to see how correctly the newcomers adhered to the direction indicated. An example will show how that can be done and what comes out of it: an observation hive was set up in a large level field. Just outside the flight entrance we enticed a few bees to a feeding dish and with them walked slowly 600 meters eastward. During the trip the sweetness of the sugar solution was so slight that the (individually marked) foragers did not dance and accordingly no newcomers appeared. No scent was added at the feeding place. When we had reached the goal, scent plates with a little lavender fragrance were distributed on the ground 50 meters nearer the hive and fan-wise at both sides, but without food; at the feeding place the weak sugar solution was replaced with sweeter material and at the same time some lavender scent was added. The foragers that came home now danced and showed the direction of the goal. Newcomers that had been aroused flew out and went hunting for the scent they had perceived on the dancers. When they came en route close to a plate with lavender scent they were attracted to it. From the number of flights to the several scent plates we can see what direction they are hunting in. The numbers beside the scent plates (Figure 53) specify how

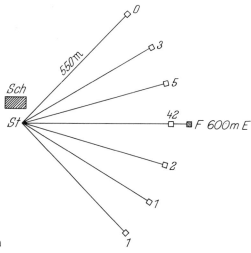

Figure 53. Example of a fan-shaped experiment. *St*, beehive; *Sch*, shed; *F*, feeding place with lavender scent, 600 meters from the hive. The numbers record the number of visits by newcomers to the 7 food-free scent plates, that were set out 550 meters from the hive at angular intervals of 15 degrees. (From von Frisch, *Tanzsprache und Orientierung der Bienen* [Springer-Verlag, 1965], p. 159.)

many flights came to each during the first 50 minutes of the experiment. Such "fan-shaped experiments" show that the direction pointed out is adhered to quite precisely.

In "step-wise experiments" we set out all the scent plates in the direction of the feeding place but at graded distances from near the hive to far beyond the feeding place. Then the newcomers flew off over and beyond the nearby scent plates; they also paid no attention to the scented bait at too great an interval from the hive, but instead stubbornly scoured the neighborhood approximately at the indicated distance (von Frisch, 1967, pp. 156–161, 84–96).

If a feeding place is near the hive it is found easily by the newcomers. We know that under such conditions the foragers perform round dances (p. 72). The latter do not provide any precise description of the location, they merely express the symbolic summons to investigate thoroughly the surroundings of the hive. Experimentally it has been shown that the round dances are understood in this sense.

We first set out a scented feeding place 10 meters east of the observation hive. Then cards with the same scent but without food were placed on the ground 25 meters from the hive, one in the same direction and others to the north, west, and south (Figure 54). The numbered bees returning from the nearer feeding place performed round dances as usual; within one hour we had counted the following numbers of bees at the four scented cards lying 25 meters from the hive: to the east, 27 bees, and in the other directions, 37, 20, and 19 bees respectively (Figure 55, *left*). Thus the round dances did not indicate direction to any significant degree; the bees sought for food on all sides of the hive. Later, the same feeding vessel was moved to a point 100 meters north of the hive

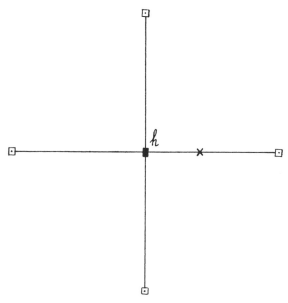

Figure 54. *h*, beehive; *x*, feeding place 10 meters east of the hive. The four squares indicate the position of cards provided with the same scent as the feeding place, but without food. See Figure 55.

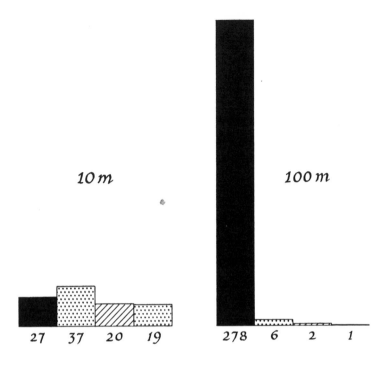

27   37   20   19   278   6   2   1

Figure 55. Bees collecting food at a distance of 10 meters perform round dances and thus send out other bees in about equal
numbers in all directions (*left*). Bees collecting food at a distance
of 100 meters perform tail-wagging dances and send out the other
bees in the correct direction (*right*). The heights of the columns
show the number of visitors at the 4 cards (see Figure 54), the
black column indicating the card in the direction of the feeding
place, the column with oblique lines indicating the card in the
opposite direction, and the dotted columns indicating the cards
situated at right angles to the direction to the feeding place. The
exact number of visitors is given by the figures below the columns.

and the experiment was repeated. Within one hour we now counted 278 bees visiting the scented card north of the hive, and only 6, 2, and 1 at three other cards lying 100 meters to the east, south, and west (Figure 55, *right*). The bees returning from 100 meters performed wagging dances that conveyed a clear message about the direction in which food was to be found.[1]

Figure 56. To reach the feeding place (*f*), the bees coming from the hive (*h*) have to fly around the corner of a steep hill. The hill is shown in contour.

In the mountains of my native country the terrain itself suggested the experiment of setting the feeding place behind a ridge, so that the bees would be obliged to fly a detour in order to reach it (Figure 56). I was very curious to see whether they would indicate the direction of the first or the second portion of the route. But the dancing bees actually indicated neither the first nor the sec-

[1] In this connection there are differences among different races of bees. We worked mainly with the Carniolan race. Italian bees (with yellow abdominal segments), that constitute the great majority of bees kept in the United States, begin to indicate distance when the goal is no more than 10 meters from the hive.

ond direction, but the "bee line" straight from hive to food. On one occasion I found a location on a certain mountain which was especially suitable for experiments in which the bees were obliged to reach the food by an

Figure 57. Another experiment in which the bees had to fly a detour. The feeding place is on the left side of the picture (white cross), and the hive is about at the same level on the other side, behind the ridge.

extensive detour. The observation hive was on one side of a steep ridge and the feeding place was on the other side (Figure 57). We gradually moved the feeding vessel around the ridge while a group of bees continued to gather sugar-water from it. But the result was a disappointment, for the bees surprised us by flying up and over the ridge. They had apparently discovered very quickly that the flight path over the rock was a little shorter than the detour around the end of the ridge. They did not deviate from the compass direction, but the actual distance of their flight was much greater than a straight line drawn through the ridge from hive to feeding place. By timing the dances in the observation hive we could see that they were indicating the distance of the actual flight over the ridge rather than straight-line distance.

In a later year we repeated the experiment, but this time the hairpin curve from the observation hive to the feeding place was shorter than the route over the ridge. The foragers kept flying to the goal and back only via the detour on which we had initially led them around the obstacle. But in dancing they invariably indicated the air line to the goal, over which they had never flown. Thus they demonstrated very clearly to us their capacity for integrating, from the detour flown over, the direction of the goal (Figure 58).

And what do the newcomers do if they obey the

dancers' information and run into the cliff when they
go flying out? We investigated that in a later experiment. We wished to learn whether they would deviate
at random to right or left. But they did neither the one
nor the other. They held true to the direction toward
the goal that had been proclaimed to them, flew up the
face of the cliff and, crossing over the ridge, reached

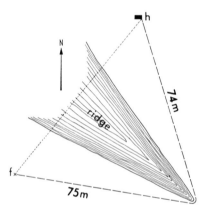

Figure 58. Detour experiment. *h*, observation hive; *f*, feeding
place; continuous line, flight path of the foragers; dashed line,
direction to the feeding place as shown by the dancers. Schematic.

the feeding place. Once this place was familiar to them
they soon discovered the shorter roundabout route over
which the foragers were flying (von Frisch, 1967, pp.
173–186).

   As early as in 1923 I thought I could understand the
language of bees. But later experiments brought many

additional surprises. One of these was the discovery that the wagging dance points out the direction of and distance to the goal. Initially that struck me as being fantastic beyond belief.

I wondered whether perhaps the bees of my observation hive had developed into a sort of scientific bee. I decided to see whether the same dances would occur in an ordinary hive. This was clearly so, for I could lift one honeycomb from a typical hive and see the same type of dances still going on, despite the disturbance, as I held the honeycomb in my hands. On one occasion as I was holding such a honeycomb covered with bees I became curious to see what would happen if I held the comb horizontal instead of vertical. The bees continued to dance, yet they could no longer orient their dances relative to gravity, for on a horizontal surface there is no up or down. Now the bees pointed the straight portion of their dances *directly toward the feeding place.* If the honeycomb was slowly rotated like a turntable they kept pointing in the direction of the feeding place, just like a compass needle. To study this phenomenon more carefully I used my observation hive instead of a single honeycomb lifted from an ordinary beehive. But now I laid the observation hive down on its side. This did not seem to disturb the bees, even if the hive was kept horizontal for many days. In the horizontal hive the dances did not change their direction with the position of the sun; they always indicated the direction straight to the feeding place.

The simplest explanation of this behavior would be to assume that the dancers still indicated the direction to the feeding place in relation to the sun—but in this case directly, and not with reference to gravity. If during the straight portion of the wagging dance a bee kept herself oriented at the same angle to the sun's rays as she had kept in flying out from the hive to the feeding place, then on a horizontal surface she would point her dances directly toward the food (Figure 59). This explanation

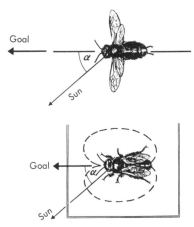

Figure 59. The principle whereby direction is indicated by means of dancing on a horizontal surface. During the wagging run (*bottom*) the bee takes such a position that she sees the sun at the same angle *a* as during her previous flight to the feeding place (*top*). (From von Frisch, *Tanzsprache und Orientierung der Bienen* [Springer-Verlag, 1965], p. 132.)

assumes that the dancing bee can see where the sun stands in the sky. But in my experiments the observation hive was always shaded by a roof. Evidently something more complicated was involved.

I wished first to see whether sunlight was necessary at all for orientation of the dances on a horizontal honeycomb. I built a movable chamber to enclose the observation hive and the observer as well (Figure 60). Inside this hut the dances could be observed by red light, which the bees cannot see, or by diffuse white light. In either

Figure 60. The observation hive enclosed in the opaque chamber. On the bench is the entrance for the bees.

case the dances were as vigorous as before, but entirely disoriented. They continually shifted in direction, and the newly aroused bees, thoroughly confused, searched for food in all directions. Evidently the direction cannot be indicated on a horizontal surface in diffuse light or in the dark. Yet in the normal hive the dances are performed in darkness—but they are then executed on a vertical surface and are oriented with reference to gravity. The horizontal dances can also be understood in terms of the normal behavior of a colony of honeybees. Dancing bees can often be seen in warm weather on the horizontal board just in front of the entrance to a hive, when some foraging bees stop at the entrance and deliver nectar to other bees that are waiting outside. In this case they dance in daylight and indicate the direction straight toward the feeding place. Inside the dark hive there is no need for such orientation without reference to gravity because there are ordinarily no horizontal surfaces suitable for dancing.

But from the point of view of the physiologist a question arises: How is it possible for bees to indicate the direction of the feeding place if they are dancing on a horizontal surface and if they cannot see the sun?

I have mentioned that when the observation hive was enclosed in a hut with opaque walls the bees could no longer indicate the direction on horizontal surfaces. My next experiment was to remove one wall of this hut and allow the bees to see only an area of blue sky at some distance from the position of the sun. When the blue sky became visible in this way the dances were instantly put in order; the bees once more pointed directly toward the food. To achieve this result it was enough to provide a crack in the wall which was only 4 inches in width. Indeed, the bees could indicate the correct direction even when we allowed them to see the sky only through a

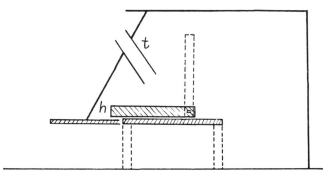

Figure 61. Cross section through the chamber, showing the bee-hive (*h*) in a horizontal position (normal position indicated by dotted lines). From the location of the dancing bees a spot of blue sky can be seen through the tube (*t*).

tube 16 inches long and 6 inches in diameter (Figure 61). If we closed the tube, the dances were disoriented; if we opened it again, they were correct.

In these experiments the tube was pointed upward toward the north sky. The straight part of the dances was directed toward a feeding place to the west. I next placed a mirror just outside of the tube, so that the bees still saw a blue spot, but this was now a reflected image of the sky to the south. Under the influence of the mirror the direction of the dances shifted; the bees pointed east instead of west. From this experiment it was clear that bees can perceive in the sky some phenomenon dependent on the sun's position even though they cannot see the sun directly. As soon as they have recognized the position of the sun they can indicate the direction to the feeding place, keeping the same angle to the sun as they would keep if they were outside the hive and flying to the source of food.

Sometimes a cloud would pass across the area of sky visible through the tube; when this happened the dances became disoriented, and the bees were unable to indicate the direction to the feeding place. Whatever phenomenon in the blue sky served to orient the dances, this experiment showed that it was seriously disturbed if the blue sky was covered by a cloud. We know of such a phenomenon, which is visible in the blue sky, which is related to the position of the sun, and whose intensity is greatly reduced by clouds. It is the polarization of light (von Frisch, 1948).

Light rays coming directly from the sun consist of vibrations that occur in all directions perpendicular to the line along which the sunlight travels. But the light of the blue sky has not reached us directly from the sun; it has first been scattered from particles in the atmosphere. This diffuse, scattered light coming from the sky is partially polarized, by which we mean that more of it is vibrating in one direction than in others. Devices such as a Nicol prism or a piece of polaroid transmit only the light vibrating in one direction; for this reason they are sometimes called analyzers of polarized light. Hence if one holds such a device before one's eyes while looking at the blue sky, the sky will appear darker or lighter as the analyzer is rotated. It is easy to see by simple inspection of the blue sky through a piece of polaroid that the intensity of polarization varies with the distance from the sun. If one looks at the sky 90 degrees away from the sun, one sees the greatest change in brightness as the analyzer is rotated; as much as 70 per cent of the light may be polarized; but nearer to the sun and farther away from it the degree of polarization diminishes, and it becomes very feeble or absent altogether in the region close to the sun and around the point just opposite.

There is also another relationship between the position of the sun and the polarization of the blue light from the sky. The plane of polarization of the light from any point in the sky is always perpendicular to a second plane determined by three points: the eye of the observer, the spot of sky at which he looks, and the sun.

This rule fails to hold only in the neighborhood of the sun and close to the point of sky just opposite to it.[2]

Theoretically at least it is thus possible to determine the sun's position from inspection of an area of blue sky, provided one can detect the polarization of the light with some sort of analyzer.

But such a theoretical possibility could lead to nothing more than speculation until we had learned whether the dances of bees were influenced in any way by changing the polarization of the light coming to them from the blue sky. Such a test was first made by using a sheet of polaroid, like the visors that are sometimes attached to automobile windshields for protection against glare. I used a sheet about 6 inches wide and 12 inches long, and every part of the surface of the sheet acted as an analyzer; the light passing through it became polarized in one plane. I placed this sheet over the glass window of my observation hive while the latter lay in a horizontal position. To simplify the experimental situation I placed opaque screens on three sides of the hive; and there was also a roof to exclude light from above. The dancing bees could see the blue sky in only one direction.

Now I rotated the polaroid sheet like a turntable to

[2] These geometrical relations result from the manner in which light is scattered by particles in the atmosphere; a detailed explanation of the whole matter can be found in any standard textbook of optics; see von Frisch, 1967, pp. 381, 382.

the right or the left above the bees. Never shall I forget the joy with which I saw the dancers react to it at once and shift the line of their wagging runs in the direction of rotation. Without exception the dances pointed farther toward the right after a rotation to the right, and farther toward the left after a rotation to the left. This of itself demonstrated that they orient with reference to the polarization of the blue sky (von Frisch, 1949, 1950). But they did not always shift their indication of direction by precisely the angle through which I had rotated the polaroid sheet. For example, it sometimes happened that after a rotation of 30 degrees the line of dancing was shifted in the same direction, but by 35 degrees. In order to comprehend this we need more intimate knowledge about the polarized light in the blue vault of heaven and about its analysis by the eye of the bee.

As generally with insects, at the right and left sides of the bee's head there sits a compound eye, each of which consists of several thousand ommatidia (Figure 62). In the human eye the lens casts onto the retina an image of the environment, which is taken up by the sensory cells and transmitted to the higher centers. The compound eye functions in accordance with a different principle. The ommatidia stand side by side in a rigorous arrangement, with their axes diverging slightly (Figure 62). Hence each one looks out in a somewhat dif-

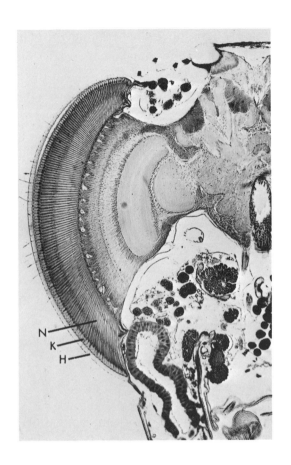

Figure 62. Longitudinal section through the eye of the bee (worker), in a dorsoventral direction. *H*, cornea; *K*, crystalline cone; *N*, retina (rhabdoms). Photograph by Agnes Langwald. (From von Frisch, *Tanzsprache und Orientierung der Bienen* [Springer-Verlag, 1965], p. 490.)

ferent direction and there views "bright" or "dark," "yellow" or "blue," whichever accords with its line of sight. These individual images of the ommatidia are pieced together into a picture of the entire environment like the small stones in a mosaic. In Figure 63 a strongly

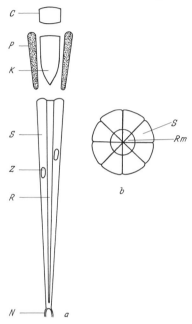

Figure 63. A single eye (ommatidium or omma) from an insect's compound eye: (*a*) longitudinal section; (*b*) cross section, at higher magnification, through the sensory cells; *S*, sensory cells; *Z*, cell nucleus; *N*, nerve fiber; *C*, corneal lens; *K*, crystalline cone; *P*, pigment. On its internal face each sense cell forms a rhabdomere, *Rm;* the rhabdomeres may unite to form a rhabdom, *R.* Schematic. (From von Frisch, *Tanzsprache und Orientierung der Bienen* [Springer-Verlag, 1965], p. 390.)

magnified ommatidium is depicted. Light enters through the cornea (C) and is conducted through the crystalline cone (K) to the sensory cells (S). The latter are connected with the nerve centers via the nerve fibers (N). Each of the eight sensory cells bears an inner border; these eight borders constitute the central visual rod (the rhabdom R). In a cross-section one sees (right-hand figure) that the sensory cells of the ommatidium are arranged in the form of a rosette. Their borders, the rhabdomeres [3] (Rm), demand our special interest. For in them the light waves are converted into nervous excitation.

Let us now recall that we ourselves can see the pattern of polarization if we look at the blue sky through a polaroid sheet and rotate the latter (p. 117). The region focused on grows brighter or darker. What is going on can be clarified with a simple experiment (Figure 64): we hold a polaroid sheet (as a model of the blue sky, from which polarized light comes) against a bright background in such a way that the plane of vibration of the transmitted light is horizontal (double-headed arrow). Now we put smaller strips of polaroid at various orientations in front of it. When the planes of vibration are at right angles, the covered area is black (*right*). The light that comes from the first sheet cannot pass through the second. As the angle between the axes of

[3] Greek: *rhabdos* = rod; *meros* = part.

vibration of the two sheets is decreased the brightness increases steadily, and reaches a maximum when the axes of the two coincide (*left*—a slight degree of darkening remains because the polaroid sheets are not totally colorless). Thus, if the axis of vibration of the first sheet is not known it can be found by means of a second whose axis of vibration is known.

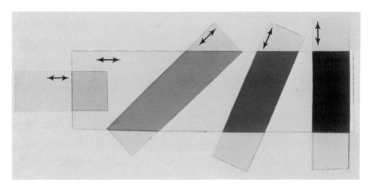

Figure 64. Polarizing sheets are placed over one another at different angles, with increasing extinction of the light. The plane of vibration of the light transmitted by the sheets is indicated by the double-headed arrows. (From von Frisch, *Tanzsprache und Orientierung der Bienen* [Springer-Verlag, 1965], p. 390.)

Bees cannot grasp a sheet of polaroid and rotate it against the blue sky. But H. Autrum had a good idea: if the radially situated rhabdomeres (*Rh* in Figure 63) were themselves to polarize light passing through them, each in a different plane in accordance with its attitude, then in theory each individual ommatidium could analyze the plane of vibration of polarized light at any point of the sky. He was able to bolster this assumption via electrophysiological experiments (Autrum and Stumpf, 1950). I tried to test his hypothesis with free-flying bees. From a polaroid sheet I cut triangles and joined them together to make an artificial bee's eye, as shown in Figure 65 (cf. Figure 63 *right* with Figure 65*b*). If this model, appropriately mounted (Figure 66), is directed at the blue sky, one obtains different pat-

Figure 65. (*a*) Pattern for cutting a polarizing sheet to make a star-shaped model. (*b*) The double-headed arrows indicate the plane of vibration of the polarized light. (From von Frisch, *Tanzsprache und Orientierung der Bienen* [Springer-Verlag, 1965], p. 391.)

terns for different parts of the sky, depending on the plane of vibration of polarized light at the observed area (Figure 67).

Now we have come so far that we can return to the dancing bees. We shall try to understand how, whenever a polaroid sheet is rotated above the bees, the observed shift in their angle of dancing comes about. From a great number of experiments I present but three examples as illustrating the method.

1. The observation hive is laid on its side. Bees marked with numbers are frequenting a feeding place 200 meters to the west. During their dances in the hive they are able to see a limited spot of blue sky in the north about 45 degrees above the horizon. Beside the hive is our model of the bee's eye. We observe the pattern of the sky in the north and then place above the dancing bees a sheet of polaroid so oriented that the

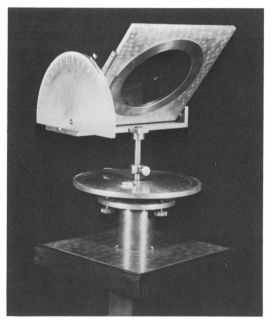

Figure 66. The star-shaped model, mounted; azimuth and inclination may be read from two graduated arcs; in the center of the model is a small mirror for monitoring the vertical direction of sighting. (From von Frisch, *Tanzsprache und Orientierung der Bienen* [Springer-Verlag, 1965], p. 391.)

plane of vibration remains unchanged for them (the
proper placement is found by putting a polaroid sheet
in front of the model and rotating it). Under such
conditions the dances remain unaltered and point directly to the feeding place.

2. From the horizontal comb the bees can see a

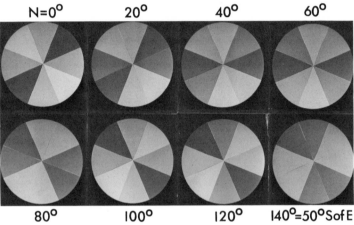

Figure 67. Photograph of the blue sky through the star-shaped
model from 3:03–3:11 P.M. September 11, 1964, near Munich. Films
exposed at 20-degree intervals from N to 50 degrees S of E, 45
degrees above the horizon. Photograph by M. Renner. (From von
Frisch, *Tanzsprache und Orientierung der Bienen* [Springer-
Verlag, 1965], p. 436.)

limited spot of blue sky in the west. Again we place
above the dancers a sheet of polaroid in such a position
that the plane of vibration of the light from the sky
remains unchanged. The dances continue to point correctly, toward the feeding place in the west. Now I

rotated the polaroid sheet 30 degrees counterclockwise; the direction of the dances changed 35 degrees and came to point southwest instead of west. I pointed the artificial eye toward the portion of the western sky which was visible to the dancing bees, and saw the brightness pattern reproduced in the left-hand photograph of Figure 68. Next I placed a sheet of polaroid in front of the instrument, in the same orientation as that lying over the observation hive. The brightness pattern

Figure 68. Photograph of blue sky through the artificial eye at 9:40–10:00 A.M., September 5, 1949. *Left:* West sky as seen through the instrument. *Center:* Same, with polaroid placed over the instrument in the same orientation as the polaroid placed over the observation hive. *Right:* The same pattern is to be seen through the instrument, *without* the polaroid sheet, at 34 degrees north of west.

shifted to that shown in the center photograph of Figure 68. If I now removed the sheet of polaroid and turned the artificial eye toward other directions, I found this same pattern at 34 degrees north of west (right-hand photograph of Figure 68). This pattern was not to be observed in any other direction.

From these results it became clear why the bees had turned south by a slightly greater angle than the 30 degrees through which the polaroid was rotated. During flight from hive to feeding place they had seen the left-hand pattern of Figure 68 directly ahead and to the west. But 34 degrees to the right they had seen the pattern shown in the center and right-hand photographs of Figure 68. When the polaroid sheet was rotated through 30 degrees before the window of the observation hive, the center pattern of Figure 68 was thereby presented to the bees as the only visible area of polarized light. These relationships are diagramed in Figure 69, where the original pattern visible straight ahead, to the west, is designated No. 1, while the pattern visible out of doors at 34 degrees north of west is called No. 2. In our experiment we first presented pattern No. 1, and the bees pointed directly toward it; but on applying the polaroid we offered, in the same window, pattern No. 2. The second pattern was that which the bees had seen at 34 degrees to their right as they had flown out from the hive to the feeding place, and hence they now pointed their dances 35 degrees to the left of the position where this pattern was presented. Thus they indicated a direction which deviated 35 degrees from the flight line to the feeding place but which conformed within one degree to the direction predicted by our hypothesis.

By means of many control experiments with the

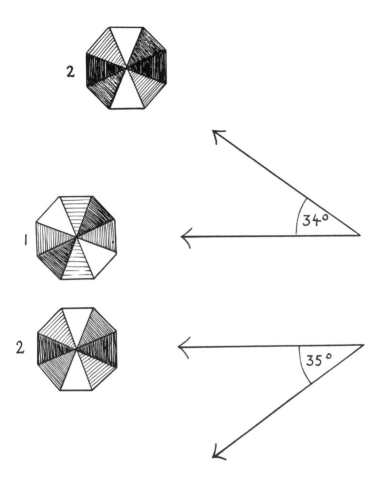

Figure 69. Upper: The bees see straight ahead of them (to the west), when flying from hive to feeding place, pattern No. 1; they also see pattern No. 2 at 34 degrees to the right. *Lower:* In the experiment the dancing bees saw pattern No. 2 as they looked out from the window of the observation hive; hence they indicated by their dances (with an error of 1 degree) that the feeding place lay 35 degrees to the left of the window.

artificial eye we showed that the direction of the dances always shifted by the same angle as that to which the brightness pattern in the sky was displaced under the particular conditions of the experiment.

3. Out of a total of 83 experiments of this kind during the summer of 1949, there were 16 in which the bees did not point in any definite direction when the polaroid was in place, but rather danced in an entirely disoriented fashion. Simultaneous observations with the artificial eye showed that in these cases—and *only* in these cases—the pattern which was produced by placing a polaroid over the instrument could not, at the time of the experiment, be found in *any part of the sky* when one looked through the instrument without the sheet of polaroid in front of it.

Through such experiments and yet others we are certain that bees are able to orient by means of reference to the polarization pattern of the blue sky. In this connection we thought at first of a special capacity of the bees. We were accustomed to exceptional accomplishments from them. But it soon became evident that other insects too could perceive the polarized light of the sky and could orient accordingly. Ants make use of the sun and of polarized light in order to maintain a constant compass direction in their journeys away from the nest and in order to find their way home again by the same route. Crustaceans and spiders that live on the seashore or on the margins of lakes and rivers orient in the same way. If a storm carries them up into dry terrain, they will set out toward the shore at any hour of the day. For, like the bees they too are familiar with the daily course of the sun and they too possess an "internal clock," without which they would be unable to use the sun and sky as a compass. From one year to another the discoveries piled up. Today it may be stated that the perception of polarized light is a capability distributed generally among insects and crustaceans, that it also occurs among spiders and even among octopuses and squids with their altogether differently constructed eyes, and that it is an important means of orientation. But among human beings, and among all vertebrates, one looks for it in vain. The polarization patterns in the sky remain a closed world to these creatures. Their eyes lack an analyzer for polarized light.

But now, just where is this analyzer in the compound eye? We have discussed the hypothesis that the radially positioned sensory cells polarize in differing planes the light that passes through them and thus, like the "artificial bee's eye," constitute an analyzer. Only with the electron microscope has more precise insight into this process been achieved.

Let us once more examine in Figure 63 the cross section through the sensory cells of an ommatidium. We know that the borders of the sensory cells, the rhabdomeres (Rm), are the light-sensitive elements. With a 25,000-fold magnification Fernandez-Moran (1956) discovered in the rhabdomeres of a fly's eye an extraordinarily orderly fine structure. In Figure 70 is repre-

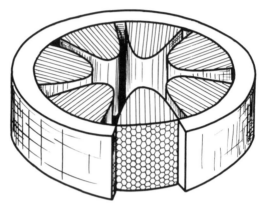

Figure 70. Cross-sectional slice cut from an ommatidium of a fly's eye, showing the radial arrangement of the fine structure in the rhabdomeres. For the sake of clarity the rhabdomeres are shown as overlarge, and the sensory cells belonging to them are merely indicated as a ring. Schematic, after sketches by Wolken, Capenos, and Turano, 1957. (From von Frisch, *Tanzsprache und Orientierung der Bienen* [Springer-Verlag, 1965], p. 428.)

sented an excised section of the rhabdomeres, while the outer portions of the sensory cells are merely indicated schematically as a ring. In the rhabdomeres (in the fly the rosette contains but seven [4]) a radial striation

[4] In some species of flies a short sensory cell with rhabdomeres has been found (Melamed and Trujillo-Cenóz, 1968).

can be seen. At still stronger magnifications one sees numerous exactly parallel tubelets ("microvilli," Figure 71). In them are laid down in oriented layers the molecules of the chemical that absorbs light and converts it into nervous excitation. In accomplishing this the light is partially polarized. Its plane of vibration is determined by the orientation of the molecules of visual substance ("dichroic absorption"). Additional experiments have demonstrated the validity of this con-

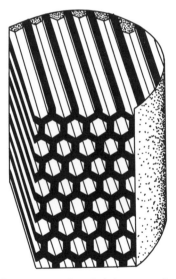

Figure 71. Small segment of a rhabdomere of a fly's eye, schematic; as determined by electron microscopy, the tiny tubelets are perpendicular to the direction of transmission of light. After Goldsmith and Philpott, 1957. (From von Frisch, *Tanzsprache und Orientierung der Bienen* [Springer-Verlag, 1965], p. 428.)

cept (Langer, 1965; Waterman and Horch, 1966; Shaw, 1966, 1967; Eguchi and Waterman, 1968). Here we may content ourselves with the observed fact that the model of the "artificial bee's eye" has been realized in Nature.

But not in the bee! She found a somewhat different way, for in her eyes the delicate tubelets always run in the same direction in each pair of adjacent rhabdomeres. Figure 72 (Goldsmith, 1962) shows this in an electron-

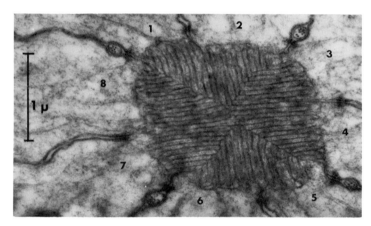

Figure 72. Cross section through a bee's ommatidium at the level of the sensory cells. Of the eight visual cells (1–8) only the inner portions with the rhabdomeres are in view. Each pair of adjacent rhabdomeres is fused together, with their tubelets (recognizable as striations) in parallel. Electronmicrograph, magnification 29,000 x. After Goldsmith, 1962. (From von Frisch, *Tanzsprache und Orientierung der Bienen* [Springer-Verlag, 1965], p. 432.)

micrograph. If in accordance with this we make the model of the artificial bee's eye four-segmented instead of eight-segmented, the view of the sky shows simpler images (Figure 73). In each ommatidium only two areas of differing brightness are formed. The difference in brightness changes with the plane of vibration of the

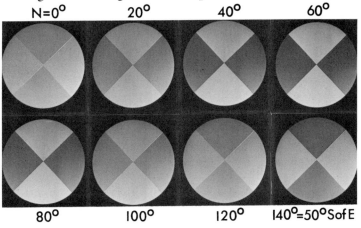

N=0°   20°   40°   60°

80°   100°   120°   140°=50°SofE

Figure 73. Photograph of the blue sky at the same hour and in the same direction as in Figure 67, but through a four-segmented star-shaped model. Photograph by M. Renner. (From von Frisch, *Tanzsprache und Orientierung der Bienen* [Springer-Verlag, 1965], p. 436.)

polarized light and thus may serve to define it. Perhaps this pattern is analyzed more easily by the brain than complex patterns like those in Figure 67. It has also been found that this crossed arrangement of the fine structure is the one realized most often: in the majority of insects, in crustaceans, and in cuttlefish.[5]

Now the bee's dances have once again led us back to the starting point of these discussions, to the bee's eye and what it does. The bees were the first to reveal to us that the polarized light of the sky can be perceived and used for orientation. Today we know that most arthropods and many molluscs possess this same ability. But only the bees have succeeded in putting polarized light to work in their system of communication. With its help they enable their comrades to find a distant goal all by themselves.

Like all of earth's living creatures, the bees have evolved their present level of organization in the course of a long ancestral history. Their language too must have developed from simpler antecedents. We know of fossil bees that were enclosed in amber fifty million years ago and that have thus been preserved; there are petrified specimens, thirty million years old, that al-

[5] With the four-segmented foil the identical pattern may occur at two different regions of the sky (for instance in Figure 73: 20 degrees east of North and 10 degrees south of East). This is ambiguous in respect to the true direction. But the bee does not view the sky through a *single* ommatidium. In the neighboring units the brightness of the sectors changes in contrary senses (cf. Figure 73). Consequently even these two regions of the sky are to be distinguished unequivocally by means of the adjacent patterns.

ready display all the bodily characteristics of the honeybee of today. But these are silent witnesses of the past. They tell us nothing about whether and how they may have communicated with one another. Relevant clues can be found, however, if today we go in search of preliminary stages of communication among their more primitive living relatives.

In Brazil there are many genera and species of wild stingless bees (meliponines). Like the honeybees they form colonies, whose social organization, however, is developed to a very different level. They are not direct ancestors of the honeybee. They constitute a side branch in the historical evolution of the line. But they have simpler forms of communication, such as might have prevailed as preliminary stages among the direct ancestors of the honeybee too.

These matters are simplest among species of the genus *Trigona*. The nest of these small bees is built in as disorderly a fashion as a bumblebee's nest. If flowers rich in nectar are discovered in the neighborhood, the finder flies home, runs excitedly about on the comb, and jostles her comrades, simultaneously emitting buzzing sounds. But nothing indicates the distance or direction of her discovery. Nonetheless she brings along a clue: the fragrance of the kind of flower she visited still clings recognizably to her body. The newcomers hunt through the neighborhood in all directions for this odor (Lindauer, 1956, 1961). Here we are probably at the root of the bees' language.

The genus *Melipona* forms larger, more highly organized colonies. Communication too has developed a step further. When the discoverer of a profitable find is running about excitedly on the comb, she emits bursts of buzzing tones, in fact short ones when the discovery site is nearby and longer ones for a greater distance. The duration of the individual buzzing tones shows the same sort of orderly relationship with increasing distance as the duration of the waggling runs in the honeybee, which are also accompanied by buzzing (see p. 93). Even in *Melipona* the comrades can learn from this the distance to the goal. Whether they give heed to these signals, and with what precision, is not yet adequately clarified. Certainly the direction is imparted in none so abstract a way. It is shown to the comrades directly by means of an escorting flight, during which the original forager traverses in a slow zigzag the first 30–50 meters away from the nest, so that a newcomer is able to follow her. But then the association is interrupted, as the forager disappears in a rapid flight to the goal. Nestmates that have been aroused return to the nest exit and continue to trail after further foraging flights, 20–30 times; then they pursue this direction independently in the search for the goal (Lindauer and

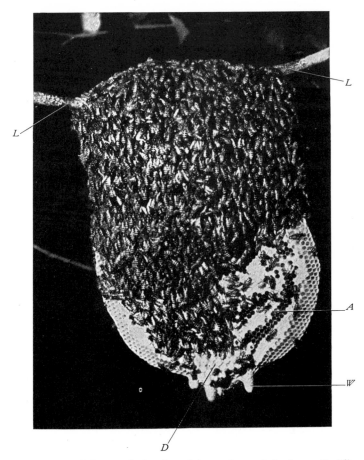

Figure 74. Colony of the dwarf honeybee, *Apis florea* F. The
bees were brushed away from the lower part of the comb in order
to show the brood nest: *A*, worker brood; *D*, drone brood; *W*,
queen cells. To the right and left of the comb the bees have applied a ring of "glue" (*L*) to protect the nest against ants; it consists of a resinous substance collected from plants. After Lindauer,
1956. (From von Frisch, *Tanzsprache und Orientierung der Bienen*
[Springer-Verlag, 1965], p. 301.)

Kerr, 1958; H. Esch, I. Esch, and W. E. Kerr, 1965; Esch, 1967). One may imagine that some meliponines —by far not all species have been studied—buzz on the comb whenever they are running in the direction of the goal. The buzzing, indeed, is produced by the flight muscles and may be a symbolic token of the outbound flight to the goal.

Figure 75. A normally occupied comb of *Apis florea*, viewed obliquely from above in order to show the horizontal dance floor. After Lindauer, 1956. (From von Frisch, *Tanzsprache und Orientierung der Bienen* [Springer-Verlag, 1965], p. 302.)

But these are mere suspicions. The most definite indications as to the origin of the language of the bees have been obtained from presently existing species of the genus *Apis*. Unfortunately there are besides our honeybee (*Apis mellifera*) only three more species, all in India. All of them perform perfect round dances and wagging dances, and thus can tell us nothing about their origin. But in one respect the most primitive among them, the dwarf honeybee (*Apis florea*), behaves differently. Only on a horizontal surface is she able to point the direction to the goal: during the waggling run she maintains the same angle with the sun as during the previous flight to the flowers. The dwarf honeybee builds her nest high in the crown of trees; a single small comb on its upper edge is expanded into a platform (Figures 74, 75). From this dance floor she can see the sun and the polarized light of the sky. Only here does she perform oriented dances. If she is forced onto the vertical lateral faces of the comb the dances are disoriented. She has not learned how to transpose into the direction of gravity. Consequently she never builds her nest in sheltered caverns. There her communications would break down. Thus, indicating direction on a horizontal surface, which initially I envisaged as an astounding accomplishment, is shown to be the original way of specifying to one's comrades the direction to the goal (Lindauer, 1956, 1961).

How did the bees on the vertical comb in the dark hive happen upon the practice of expressing the solar angle as the angle with the perpendicular? We do not know. But today this "discovery" no longer seems as incomprehensible as it did earlier. A similar transposition has been found in some other insects, where it seems without importance. An ant that is allowed to run about on a level surface in a dark room maintains toward a lamp some chosen angle, just as in the open she makes use of the sun in order not to lose her way. For example, she may be running at an angle 20 degrees to the right of the lamp (Figure 76a). If the light is now extinguished and at the same time the running surface placed in a vertical plane, she will set out on the same angle with the perpendicular that previously she had maintained with respect to the light. In doing so, however, she is not altogether precise about the direction and may run at some 20 degrees to the right or left, either in an upward or downward direction (Figure 76b). A ladybug comes closer to a bee's behavior, in that it sets the direction upward equivalent to that toward the light, but it may run 20 degrees off course, equally often to right or left (Figure 76c). The dung beetle's transposition is correct laterally, but—exactly opposite from the bee—he substitutes the direction downwards for that to the light (Figure 76d). There exist yet other variants, the only common feature of which is that the size of the angle of orientation is maintained when the guidance of motion passes over from the visual to the gravitational sense. It seems that the higher centers of the nervous system, which regulate the direction of movement, adhere steadfastly to the

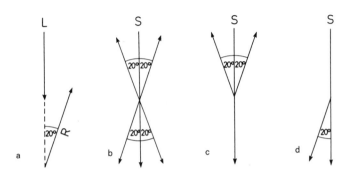

Figure 76. (*a*) A straight course *R* is maintained on a horizontal surface by keeping a constant angle—here 20 degrees to the right—relative to the light from a source *L*; (*b*) ant; (*c*) ladybug; (*d*) dung beetle exhibit different ways of transposing angles in the dark with reference to the direction of gravity *S* after the supporting surface is tilted to the vertical.

angle of orientation and accept gravity as a replacement when they lose the light as a point of reference.

On this foundation the progress made by the honeybees has been erected. They identify firmly the direction toward the sun with the direction upward, and they indicate every angular deviation to right or left unambiguously with a corresponding deviation of the waggling run. Additionally they are able in the dark hive to transpose from memory to the angle with the perpendicular the angle to the sun that they maintained during their flight to the feeding place, and to do it so accurately that their comrades succeed in finding the indicated goal by, so to speak, retranslating from the language of gravity into the language of light.

This step took place first within the genus *Apis*. It was particularly important. For it has allowed the bees to shift from settlements under the open sky to living in protected hollows in trees, cliffs and the ground and thus to maintain themselves through the winter even in unfriendly latitudes. It was a development which is largely responsible for the fact that the language of the bees is a unique accomplishment which, as far as we know at present, has no parallel in any other animal.

Bibliography

Aufsess, A. von. 1960. Geruchliche Nahorientierung der Bienen bei entomophilen und ornithophilen Blüten. *Zeitschrift für vergleichende Physiologie*, 43: 469–498.

Autrum, H. 1949. Neue Versuche zum optischen Auflösungsvermögen fliegender Insekten. *Experientia (Basel)*, *5:* 271–277.

——. 1952. Über zeitliches Auflösungsvermögen und Primärvorgänge im Insektenauge. *Die Naturwissenschaften*, 39: 290–297.

—— and H. Stumpf. 1950. Das Bienenauge als Analysator für polarisiertes Licht. *Zeitschrift für Naturforschung*, 5b: 116–122.

—— and V. von Zwehl. 1964. Die spektrale Empfindlichkeit einzelner Sehzellen des Bienenauges. *Zeitschrift für vergleichende Physiologie*, 48: 357–384.

Bauer, L. 1939. Geschmacksphysiologische Untersuchungen an Wasserkäfern. *Zeitschrift für vergleichende Physiologie*, 26: 107–120.

Beling, I. 1929. Über das Zeitgedächtnis der Bienen. *Zeitschrift für vergleichende Physiologie*, 9: 259–338.

Beutler, R. 1930. Biologisch-chemische Untersuchungen am

Nektar von Immenblumen. *Zeitschrift für vergleichende Physiologie*, 12: 72–176.

Brown, P. K. and G. Wald. 1964. Visual pigments in single rods and cones of the human retina. *Science*, 144: 45–52.

Daumer, K. 1956. Reizmetrische Untersuchung des Farbensehens der Biene. *Zeitschrift für vergleichende Physiologie*, 38: 413–478.

——. 1958. Blumenfarben wie sie die Bienen sehen. *Zeitschrift für vergleichende Physiologie*, 41: 49–110.

Eguchi, E. and T. H. Waterman. 1968. Cellular basis of polarized light perception in the spider crab *Libinia*. *Zeitschrift für Zellforschung*, 84: 87–101.

Esch, H. 1961. Über die Schallerzeugung beim Werbetanz der Honigbiene. *Zeitschrift für vergleichende Physiologie*, 45: 1–11.

——. 1964. Beiträge zum Problem der Entfernungsweisung in den Schwänzeltänzen der Honigbienen. *Zeitschrift für vergleichende Physiologie*, 48: 534–546.

——. 1967. Die Bedeutung der Lauterzeugung für die Verständigung der stachellosen Bienen. *Zeitschrift für vergleichende Physiologie*, 56: 199–220.

——, I. Esch, and W. E. Kerr. 1965. Sound: An element common to communication of stingless bees and to dances of the honeybee. *Science*, 149: 320–321.

Fernandez-Moran, H. 1956. Fine structure of the insect retinula as revealed by electron microscopy. *Nature (London)*, 177: 742–743.

Forel, A. 1910. *Das Sinnesleben der Insekten*. Munich: Reinhardt. (English translation: *The Senses of Insects*. London: Methuen, 1908.)

Frisch, K. von. 1915. Der Farbensinn und Formensinn der

Biene. *Zoologische Jahrbücher, Abteilung für allgemeine Zoologie und Physiologie*, 35: 1–182.

——. 1919. Über den Geruchssinn der Biene und seine blüten-biologische Bedeutung. *Zoologische Jahrbücher, Abteilung für allgemeine Zoologie und Physiologie*, 37: 1–238.

——. 1921. Über den Sitz des Geruchssinnes bei Insekten. *Zoologische Jahrbücher, Abteilung für allgemeine Zoologie und Physiologie*, 38: 1–68.

——. 1923. Über die "Sprache" der Bienen. *Zoologische Jahrbücher, Abteilung für allgemeine Zoologie und Physiologie*, 40: 1–186.

——. 1927. *Ein Vorschlag für die Wanderimker: Bienenzucht und Bienenforschung in Bayern*. Neumünster: Wacholtz-Verlag. Pp. 1–4.

——. 1934. Über den Geschmackssinn der Bienen. *Zeitschrift für vergleichende Physiologie*, 21: 1–156.

——. 1946a. Die "Sprache" der Bienen und ihre Nutzanwendung in der Landwirtschaft. *Experientia (Basel)*, 2: 397–404.

——. 1946b. Die Tänze der Bienen. *Österreichische Zoologische Zeitschrift*, 1: 1–48.

——. 1947. *Duftgelenkte Bienen im Dienste der Landwirtschaft und Imkerei*. Vienna: Springer-Verlag.

——. 1948. Gelöste und ungelöste Rätsel der Bienensprache. *Die Naturwissenschaften*, 35: 12–23, 38–43.

——. 1949. Die Polarisation des Himmelslichtes als orientierender Faktor bei den Tänzen der Bienen. *Experientia (Basel)*, 5: 142–148.

——. 1950. Die Sonne als Kompass im Leben der Bienen. *Experientia (Basel)*, 6: 210–221.

——. 1965. *Tanzsprache und Orientierung der Bienen*. Berlin, Heidelberg, New York: Springer-Verlag. (English translation: *Dance Language and Orientation of Bees*. Cambridge, Massachusetts: Harvard University Press, 1967.)

——. 1969. *Aus dem Leben der Bienen*. 8th ed. Berlin-Heidelberg-New York: Springer-Verlag. (English translation: *The Dancing Bees*. New York: Harcourt, Brace and World, 1966.)

Goldsmith, T. H. 1962. Fine structure of the retinulae in the compound eye of the honeybee. *Journal of Cellular Biology*, 14: 489–494.

—— and D. E. Philpott. 1957. The microstructure of the compound eye of insects. *Journal of Biophysical and Biochemical Cytology*, 3: 429–438.

Haslinger, F. 1935. Über den Geschmackssinn von *Calliphora erythrocephala* Meigen und über die Verwertung von Zuckern und Zuckeralkoholen durch diese Fliege. *Zeitschrift für vergleichende Physiologie*, 22: 614–640.

Hertz, M. 1929a; 1929b; and 1931. Die Organisation des optischen Feldes bei der Biene, I, II, und III. *Zeitschrift für vergleichende Physiologie*, 8: 693–748; 11: 107–145; 14: 629–674.

——. 1937. Versuche über das Farbensystem der Bienen. *Die Naturwissenschaften*, 25: 492–493.

——. 1939. New experiments on colour vision in bees. *Journal of Experimental Biology*, 16: 1–8.

Hess, C. von. 1913. Gesichtssinn in *Wintersteins Handbuch der vergleichenden Physiologie*. Jena: Fischer. Volume 4, pp. 555–840.

Kaissling, K. E. and M. Renner. 1968. Antennale Rezeptoren für Queen Substance und Sterzelduft bei der Honigbiene. *Zeitschrift für vergleichende Physiologie*, 59: 357–361.

Knaffl, H. 1953. Über die Flugweite und Entfernungsmeldung der Bienen. *Zeitschrift für Bienenforschung*, 2 (Zander Festschrift): 131–140.

Kühn, A. 1927. Über den Farbensinn der Bienen. *Zeitschrift für vergleichende Physiologie*, 5: 762–800.

Lacher, V. 1964. Elektrophysiologische Untersuchungen an einzelnen Rezeptoren für Geruch, Kohlendioxyd, Luftfeuchtigkeit und Temperatur auf den Antennen der Arbeitsbiene und der Drohne (*Apis mellifica* L.). *Zeitschrift für vergleichende Physiologie*, 48: 587–623.

—— and D. Schneider. 1963. Elektrophysiologischer Nachweis der Riechfunktion von Porenplatten (Sensilla placodea) auf den Antennen der Drohne und der Arbeitsbiene (*Apis mellifica* L.). *Zeitschrift für vergleichende Physiologie*, 47: 274–278.

Langer, H. 1965. Nachweis dichroitischer Absorption des Sehfarbstoffes in den Rhabdomeren des Insektenauges. *Zeitschrift für vergleichende Physiologie*, 51: 258–263.

Lex, T. 1954. Duftmale an Blüten. *Zeitschrift für vergleichende Physiologie*, 36: 212–234.

Lindauer, M. 1956. Über die Verständigung bei indischen Bienen. *Zeitschrift für vergleichende Physiologie*, 38: 521–557.

——. 1961. *Communication among Social Bees*. Cambridge, Massachusetts: Harvard University Press.

—— and W. Kerr. 1958. Die gegenseitige Verständigung bei den stachellosen Bienen. *Zeitschrift für vergleichende Physiologie*, 41: 405–434.

Maurizio, A. 1960. Bienenbotanik. In A. Büdel and E. Herold. *Biene und Bienenzucht*. Munich. Pp. 68 ff.

Melamed, J. and O. Trujillo-Cenóz. 1968. The fine structure of the central cells in the ommatidia of dipterans. *Journal Ultrastructural Research*, 21: 313–334.

Minnich, D. E. 1922a. The chemical sensitivity of the tarsi of the red admiral butterfly, Pyrameis atalanta Linn. *Journal of Experimental Zoology*, 35: 57–81.

——. 1922b. A quantitative study of tarsal sensitivity to solutions of saccharose, in the red admiral butterfly, Pyrameis atalanta Linn. *Journal of Experimental Zoology*, 36: 445–457.

——. 1929. The chemical sensitivity of the legs of the blowfly, Calliphora vomitoria Linn., to various sugars. *Zeitschrift für vergleichende Physiologie*, 11: 1–55.

——. 1932. The contact chemoreceptors of the honey bee, Apis mellifera Linn. *Journal of Experimental Zoology*, 61: 375–393.

Oettingen-Spielberg, T. zu. 1949. Über das Wesen der Suchbiene. *Zeitschrift für vergleichende Physiologie*, 31: 454–489.

Park, O. W. 1928. Studies on the sugar concentration of the nectar of various plants. *Report of the Iowa State Apiarist*. Pp. 80–89.

Porsch, O. 1931. Grellrot als Vogelblumenfarbe. *Biologia generalis (Vienna)*, 7: 647–674.

Renner, M. 1957. Neue Versuche über den Zeitsinn der Honigbiene. *Zeitschrift für vergleichende Physiologie*, 40: 85–118.

——. 1959. Über ein weiteres Versetzungsexperiment zur Analyse des Zeitsinnes und der Sonnenorientierung der Honigbiene. *Zeitschrift für vergleichende Physiologie*, 42: 449–483.

Rhein, W. von. 1957. Über die Duftlenkung der Bienen zur

Steigerung der Samenerträge des Rotklees *Trifolium pratense* L. *Zeitschrift für Acker- und Pflanzenbau,* 103: 273–314.

Ritter, E. 1936. Untersuchungen über den chemischen Sinn beim schwarzen Kolbenwasserkäfer, *Hydrous piceus. Zeitschrift für vergleichende Physiologie,* 23: 543–568.

Santschi, F. 1911. Observations et remarques critiques sur le mécanisme de l'orientation chez les fourmis. *Revue Suisse de Zoologie,* 19: 303–338.

——. 1923. L'orientation sidérale des fourmis et quelques considérations sur leurs differentes possibilités d'orientation. *Mémoires de la Societé Vaudoise des Sciences Naturelles,* 1: 117–176.

Schaller, A. 1926. Sinnesphysiologische und psychologische Untersuchungen an Wasserkäfern und Fischen. *Zeitschrift für vergleichende Physiologie,* 4: 370–464.

Scharrer, E. 1935. Die Empfindlichkeit der freien Flossenstrahlen des Knurrhahns (*Trigla*) für chemische Reize. *Zeitschrift für vergleichende Physiologie,* 22: 145–154.

Schneider, D. 1969. Insect Olfaction: Deciphering System for Chemical Messages. *Science,* 163: 1031–1037.

Scholze, E., H. Pichler, and H. Heran. 1964. Zur Entfernungsschätzung der Bienen nach dem Kraftaufwand. *Die Naturwissenschaften,* 51: 69–70.

Shaw, S. R. 1966. Polarized light responses from crab retinula cells. *Nature (London),* 211: 92–93.

——. 1967. Simultaneous recording from two cells in the locust retina. *Zeitschrift für vergleichende Physiologie,* 55: 183–194.

Sprengel, C. K. 1793. *Das entdeckte Geheimnis der Natur im Bau und in der Befruchtung der Blumen.* Berlin. (Republished in four volumes edited by Paul Knuth as Nos. 48–51 of Ostwald's Klassiker der exakten Wissenschaften. Leipzig: Engelmann, 1894.)

Steinhoff, H. 1948. Untersuchungen über die Haftfähigkeit von Duftstoffen am Bienenkörper. *Zeitschrift für vergleichende Physiologie*, 31: 38–57.

Thorpe, W. H. 1949. Orientation and methods of communication of the honeybee and its sensitivity to the polarization of the light. *Nature (London)*, 164: 11–14.

Wahl, O. 1937. Untersuchungen über ein geeignetes Vergällungsmittel für Bienenzucker. *Zeitschrift für vergleichende Physiologie*, 24: 116–142.

Waterman, T. H. and K. W. Horch. 1966. Mechanism of polarized light perception: Receptor potentials in a crab eye support a proposed two-channel intraretinal system of wide occurrence. *Science*, 154: 467–475.

Wenner, A. M. 1962. Sound production during the waggle dance of the honey bee. *Animal Behaviour*, 10: 79–95.

Wolf, E. 1927. Über das Heimkehrvermögen der Bienen. *Zeitschrift für vergleichende Physiologie*, 6: 221–254.

Wolken, J. J., J. Capenos, and A. Turano. 1957. Photoreceptor structures. III. Drosophila melanogaster. *Journal of Biophysical and Biochemical Cytology*, 3: 441–448.

Index